Rising Sun, Gathering Winds: Policies to Stabilize the Climate and Strengthen Economies

CHRISTOPHER FLAVIN

AND

SETH DUNN

Jane A. Peterson, *Editor*

WORLDWATCH PAPER 138
November 1997

THE WORLDWATCH INSTITUTE is an independent, nonprofit environmental research organization in Washington, DC. Its mission is to foster a sustainable society in which human needs are met in ways that do not threaten the health of the natural environment or future generations. To this end, the Institute conducts interdisciplinary research on emerging global issues, the results of which are published and disseminated to decisionmakers and the media.

FINANCIAL SUPPORT for the Institute is provided by the Nathan Cummings Foundation, the Geraldine R. Dodge Foundation, The Ford Foundation, the Foundation for Ecology and Development, The William and Flora Hewlett Foundation, W. Alton Jones Foundation, John D. and Catherine T. MacArthur Foundation, Charles Stewart Mott Foundation, The Curtis and Edith Munson Foundation, The Pew Charitable Trusts, Rasmussen Foundation, Rockefeller Brothers Fund, Rockefeller Financial Services, Summit Foundation, Surdna Foundation, Turner Foundation, U.N. Population Fund, Wallace Genetic Foundation, Wallace Global Fund, Weeden Foundation, and the Winslow Foundation.

THE WORLDWATCH PAPERS provide in-depth, quantitative and qualitative analysis of the major issues affecting prospects for a sustainable society. The Papers are written by members of the Worldwatch Institute research staff and reviewed by experts in the field. Published in five languages, they have been used as concise and authoritative references by governments, nongovernmental organizations, and educational institutions worldwide. For a partial list of available Papers, see back pages.

Library of Congress Catalog Number 97-061583
ISBN 1-878071-40-8

Printed on 100-percent non-chlorine bleached, partially recycled paper.

Table of Contents

ACKNOWLEDGMENTS: We wish to thank Dean Anderson, Ben DeAngelo, Michael Grubb, Florentin Krause, and Sascha Muller-Kraenner for their invaluable comments on preliminary drafts of this paper. We are also grateful to our Worldwatch colleagues Lori Ann Baldwin, Mary Caron, Suzanne Clift, Liz Doherty, Laura Malinowski, Molly O'Meara, Jim Perry, and Amy Warehime for helping us with the research, editing, and production process. Finally, we are indebted to Jane Peterson, our editor, for carefully and expertly shepherding the paper through its final stages.

CHRISTOPHER FLAVIN is Senior Vice President at the Worldwatch Institute. His research and writing focus on international energy and climate policy. He is coauthor of *Power Surge: Guide to the Coming Energy Revolution* (New York: W.W. Norton & Company, 1994). He has participated in many United Nations conferences dealing with climate issues, including the Earth Summit in Rio de Janeiro in 1992 and the first Conference of the Parties to the Framework Convention on Climate Change in Berlin in 1995. He serves on the board of the Business Council for Sustainable Energy and the American Wind Energy Association.

SETH DUNN is a Staff Researcher at the Worldwatch Institute, where he studies climate change, energy, and transportation issues. He is coauthor of the Institute's *Vital Signs 1997* report and contributes regularly to *World Watch* magazine. He participated in the first two Conferences of the Parties to the Framework Convention on Climate Change in Berlin and Geneva in 1995 and 1996.

Introduction

As environmental diplomats gather for the Third Conference of the Parties to the Climate Convention in Kyoto, Japan, in December 1997, they confront a "koan" or paradox somewhat like those contemplated by the Zen Buddhist monks seeking enlightenment in the temples of this ancient city: Although virtually all world leaders agree that rapid climate change is one of the greatest threats facing humanity, their nations have done relatively little to slow—let alone reverse—the upward trajectory of greenhouse gas emissions in the five years since the climate treaty was signed.

Amid repeated scientific warnings about the potential for greater storm intensity, rising seas, drier cropland, dying forests and coral reefs, and proliferating diseases, many governments remain convinced that any serious effort to reduce the use of fossil fuels or to slow deforestation—the chief contributors to climate change—would disrupt the economy. Reinforcing this worry, many companies and some labor unions that fear the economic consequences of efforts to slow climate change have taken to the airwaves and the corridors of power to dissuade their governments from signing a strong agreement.

The pessimism underpinning this reluctance to take on climate change may be misplaced, however. For although "koans" are not supposed to have solutions, this one does: Economical alternatives to today's greenhouse-gas-dependent technologies, industries, and lifestyles are available, and if economically beneficial policies for encouraging their adoption can be devised, the world may yet find a way of stabilizing the climate before the climate destabilizes the economy.

For such a transformation to occur, the countries represented in Kyoto must move swiftly. In the struggle to slow climate change, we have already lost the first battles. Global emissions of carbon—which in the atmosphere forms carbon dioxide (CO_2), the greenhouse gas responsible for an estimated 64 percent of the warming now under way—reached a record high of 6.2 billion tons in 1996, and have now increased nearly fourfold since 1950. This emissions binge is a planetary experiment unlike any that humanity has ever attempted, overwhelming the natural cycling of carbon by oceans and forests, and building carbon dioxide concentrations in the atmosphere to nearly 30 percent above pre-industrial levels—higher than at any time in the last 160,000 years.[1]

The Intergovernmental Panel on Climate Change (IPCC), which in 1995 confirmed a "discernible human influence on global climate," estimates that a doubling of CO_2 concentrations—bound to occur before the year 2100 on the current path—would increase the average temperature of the atmosphere at the Earth's surface by 1.0 to 3.5 degrees Celsius. This rate of warming, faster than any during the last 10,000 years, would place the world in danger of a wide range of potential dislocations. In the words of Thomas Karl, a Senior Scientist with the U.S. National Oceanic and Atmospheric Administration, altering the very composition of the atmosphere is "a momentous but rather unwanted accomplishment."[2]

The complexity of the Earth's climate system makes it impossible to predict in detail how local weather conditions across the globe will be affected. However, thousands of scientists, who have labored for years to better understand the dynamics of climate change, have discovered some alarming possibilities. Flooded cities, diminished food production, and increased storm damage all seem likely—and could well produce catastrophic economic consequences. And these changes could take many centuries to reverse, affecting the lives of billions of people—many of them not yet in a position to vote on whether they deem this a wise risk.[3]

Just as homeowners buy insurance to cope with potential dangers without knowing exactly how the future will unfold, so would governments be wise to invest in affordable means of slowing growth in greenhouse gas emissions in order to minimize the risks of climate change. The need for such insurance was formally recognized in the Framework Convention on Climate Change signed in Rio de Janeiro in 1992. But the treaty, while strong in principle—calling for stabilization of the atmosphere "at a level that would prevent dangerous anthropogenic interference with the climate system"—is weak in follow-through, specifying only reporting requirements and voluntary emissions goals. Today, fewer than half the industrial countries that were supposed to hold their year 2000 emissions to the 1990 level are on course to do so.[4]

But this failure should not distract us from the fact that many practical, economical solutions are now at hand. For as the risks of climate change have grown during the 1990s, so have the options for responding to them. During the past few years, a host of promising new technologies have moved quietly but decisively from experimental curiosity to commercial reality—allowing rapid improvement in the efficiency of energy use, and economically turning sunlight, wind, and plant matter into electricity and other useful forms of energy. These advances open up an intriguing possibility: just as the economic miracles of the twentieth century were powered by fossil fuels, the twenty-first century may be marked by an equally dramatic move *away* from fossil fuels—and from the global environmental threats they brought with them.[5]

Such a sweeping change in the world's energy system will unfold as rapidly as it must *only* if government policies—many of which currently shore up the status quo and actively retard the development of alternatives—are transformed. In the past few years, a number of governments have already demonstrated that bold, innovative policies can spur rapid adoption of the new technologies—and subsequent declines in the combustion of fossil fuels. Efforts to cut fossil fuel subsidies, improve energy efficiency standards, and provide incentives for installing renewable energy technologies and increasing forest cover are among the initiatives that have begun to alter emissions trends. In some countries, emissions have already dropped, while in others, the new measures have not yet been in place long enough to see their full effect. In still others, emissions cuts in one sector have been offset by increases in another.

In our review of the new policy measures adopted in the major industrial nations—and a few developing nations—during the past seven years, it is clear that greenhouse gas trends can be turned around with surprisingly modest shifts in policy, and that these new policies can actually boost economic development. Indeed, if all nations were simply to take up the most effective policies that have *already* been adopted piecemeal by one or more countries, global greenhouse gas emissions might now be headed downward, and national economies strengthened. Denmark, the Netherlands, and Germany, each of which has formulated a detailed strategy over the past decade, have the best track records so far—though even these nations are not doing all they could.

The fear of many industries—and of the economic ministries of many governments—that only a massive increase in energy taxes, or Byzantine regulatory regimes can slow the growth in emissions is not borne out by our survey. Until now, the countries that have achieved the most progress are the ones that have done many small things right—forging an integrated package of mutually supportive policies that bring market forces to bear to solve the climate problem. Many countries have found that when effective policies are combined, they have a greater impact than if used in isolation. Gradually increased taxes on energy or carbon are an important part of the solution, but they alone cannot remove the market barriers that stand in the way of cleaner energy technologies, more balanced transportation systems, and more stable forests.

Speaking in London in June 1997, British Environment Minister Michael Meacher, whose nation has one of the stronger records on climate change, summed up the next steps to be taken: "Energy policies should be designed to promote cleaner, more efficient energy use and production. We want a new and strong drive to develop renewable energy sources such as wind and solar energy . . . Doing so will create jobs, win exports and protect the environment." The task for governments is to follow this model. The task for the climate negotiators who meet in Kyoto this fall is to spur governments to action.[6]

This challenge is made more urgent by the enormous time pressures that world leaders now face. Each year, emissions grow, and the rate of reduction needed to stabilize the atmosphere in the next century steepens. Moreover, the world economy, led by the developing countries, is now booming, pushing greenhouse gas emissions ever higher. The world is poised to build hundreds of fossil-fuel-burning power plants, millions of kilometers of roads, tens of millions of cars, and millions of buildings with fossil-fuel-consuming appliances, heating, and air conditioning in the years ahead. Most of these structures will last decades, and replacing them would be expensive. Choosing not to build them and instead creating a new generation of less-polluting energy and transportation systems would be far more economical.[7]

Up The Emissions Mountain

The threats that rapid climate change poses to society and the natural world are unprecedented. The temperature change resulting from the projected doubling of effective CO_2 concentrations in the next century could shift the ecological zones of one third of the world's forests northward at rates well beyond the maximum migration rates of most trees and shrubs, resulting in widespread forest death and wildfires. An increase in extreme weather, such as droughts and floods, is also likely to exacerbate water and food scarcity. At the same time, a 15-95 centimeter rise in sea level could turn people now living on islands and in coastal areas into environmental refugees. And the increase of heat waves and expanded range of infectious diseases, such as malaria and cholera, could cause health problems for millions.[8]

Although a small number of credentialed skeptics have gained media attention for their hypothesis that the damage from climate change will be less than the models indicate, there is an equal or greater chance that it will actually be greater. The larger the buildup of CO_2, the higher the risk of "surprise" feedbacks such as an abrupt shift in ocean currents or a sudden release of additional greenhouse gases from dying forests or heated tundra. And according to a model devised by Jerry Malhman of Princeton University, we are on course for eventually quadrupling, not just doubling, greenhouse gas concentrations, which could raise global temperatures by as much as 10-14 degrees Celsius over the next several hundred years. Such an increase would likely halve soil moisture levels in North America, threatening harvests in the world's "breadbasket." It would also raise sea levels as much as two meters and, by reducing the water salinity that drives oceanic circulation, eventually bring the ocean's heat-carrying conveyor belt to a halt—which could cause a dramatic shift in northern climates. In an August 1997 article in *Nature,* Swiss scientists suggest that this decreased circulation would carry less carbon into the deep ocean where much of it is stored, further adding to warming.[9]

The central goal of the Framework Convention on Climate Change—stabilizing the climate before it disrupts human society and the natural world—depends on slowing the rate of warming to a maximum of 0.1 degrees Celsius per decade, according to scientific estimates. In an article published in *Science* in June 1997, Swedish atmospheric scientists estimate that staying within the maximum rates of change that have prevailed for the past 200,000 years would mean holding the concentration of carbon dioxide to less than 400 parts per million (ppm). This compares to the current level of 362 ppm, which is rising at a rate of about 15 ppm per decade.[10]

To bring atmospheric concentrations within this "safe landing corridor" will require a steady descent from the precipitous greenhouse gas emissions mountain we have climbed over the last century: cutting greenhouse gas emissions to an average of less than half the current level during the next century. But emissions are headed in the other direction. (See Figure 1.) Between 1990 and 1996, worldwide emissions of carbon from the burning of fossil fuels rose 5 percent, or the equivalent of 300 million tons each year. The higher they go, the steeper the ultimate reductions will have to be, and the greater the chance of associated economic disruptions.

FIGURE 1

World Carbon Emissions from Fossil Fuel Burning by Economic Region, 1950–96

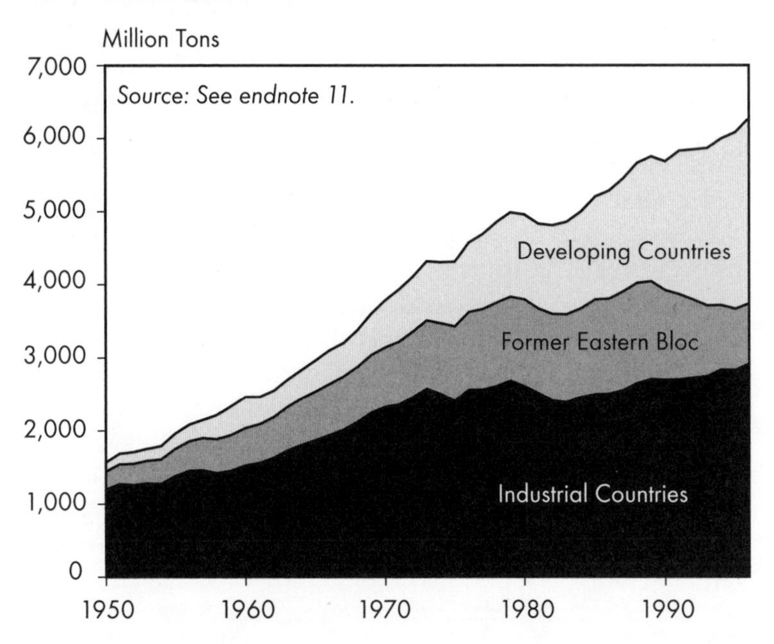

FIGURE 2

Carbon Emissions Trends, Selected Industrial and Former Eastern Bloc Countries, 1990–96

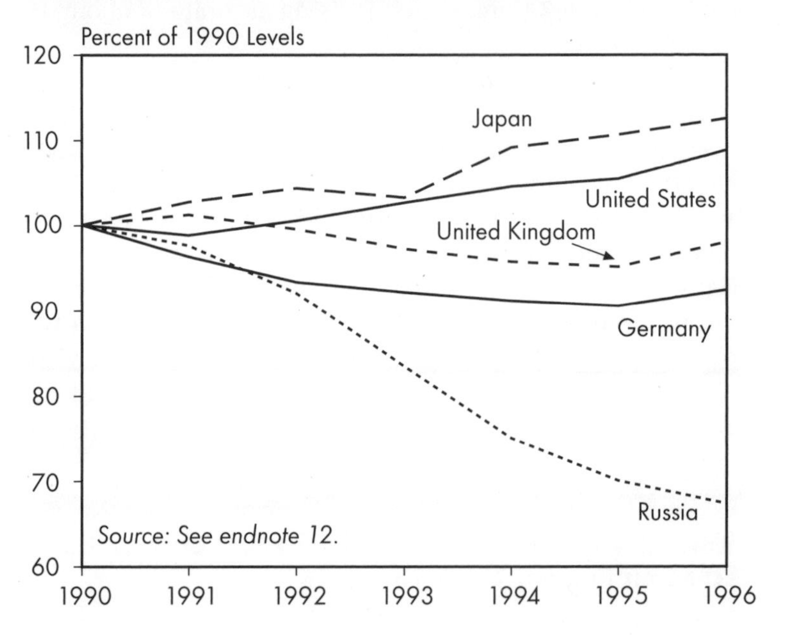

Because of the long life of much of the technology that produces greenhouse gas emissions, and the inevitable time lags in policy implementation, it is essential that the reductions begin soon. As Stephen Schneider and Lawrence Goulder of Stanford University write in the September 1997 issue of *Nature*, "It is vital to have a short-term abatement policy to bring about low-cost reductions in CO_2 emissions, even when most of those reductions will occur in the distant future."[11]

Industrial countries are responsible for 76 percent of the world's cumulative carbon emissions since 1950. Recognizing this disproportionate burden, signatories to the 1992 U.N. Framework Convention on Climate Change agreed that industrial countries should take the lead in combating climate change by drawing up and implementing national climate action plans, providing financial and tech-

FIGURE 3

Carbon Emissions Trends, Selected Developing Countries, 1990–96

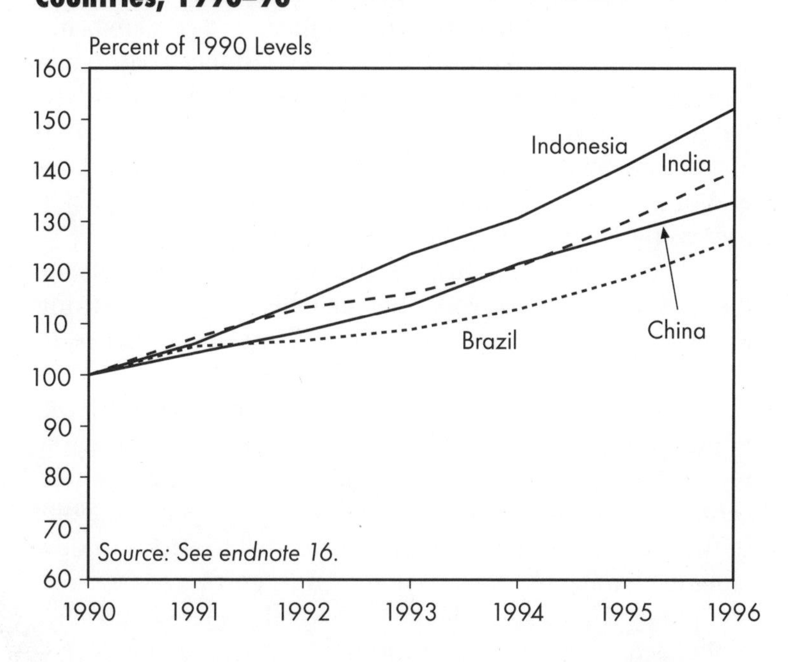

nical assistance to developing countries, and voluntarily holding emissions to their 1990 levels—or less—by the year 2000. But this goal has disappeared in the cloud of greenhouse gases belching from the automobiles and smokestacks of industrial countries. Furthest off track among the leading emitters are the United States, Australia, and Japan, whose carbon emissions in 1996 were 8.8, 9.6, and 12.5 percent above 1990 levels. In the United States, the *increase* in emissions between 1990 and 1996 is above the *total* combined annual emissions of Brazil and Indonesia, two of the largest developing countries. (See Figures 2 and 3.)[12]

The chief culprit in this recent growth in carbon emissions is transportation, which already accounts for more than 20 percent of the total resulting from energy use. The world automobile fleet surged from 50 million in 1950 to

500 million in 1997, and is projected to double over the next quarter-century as millions of people in developing countries purchase cars for the first time. In the industrial world, meanwhile, fuel prices have fallen, and cars are being sold in larger sizes and driven greater distances with each passing year, overwhelming improvements in fuel economy. Progress in raising efficiency has slowed in the residential and commercial sectors as well, where the popularity of larger homes with ever more electrical appliances is quickly increasing energy use and carbon emissions.[13]

The record in industrial countries is not universally bleak, however. The world has gained a bit of breathing room from the collapse of energy-intensive industries in eastern Europe and the former Soviet Union, driving Russia's carbon emissions down 33 percent between 1990 and 1996, while Ukraine's plummeted 56 percent. Without these reductions, global emissions would have increased 12 percent above 1990 levels by 1996, rather than 5 percent. None of these sharp declines were directly related to efforts to slow climate change, but they were driven in part by some sensible policy changes—reducing the costly energy subsidies embedded in the region's centrally planned economic systems. However, economic recovery may soon end this respite in emissions growth, as suggested by the 2.2 percent rise in Poland's carbon emissions in 1996.[14]

A few other industrial countries have had some success in limiting their greenhouse gas emissions. Carbon emissions in the United Kingdom fell roughly 2.0 percent between 1990 and 1996, while those in France declined 1.1 percent and Germany's dropped by 7.6 percent. These minor success stories are driven by a combination of new policy initiatives and broader economic trends, some predating the climate policymaking process. The United Kingdom, for instance, has reduced its coal subsidies in recent years, and France has invested heavily in nuclear power, which has largely eliminated the use of fossil fuels for electricity generation but which poses its own economic and environmental problems. Germany, meanwhile, has benefited from a combination of modest energy policy reforms and the forced shutdown of inefficient, coal-based industries and power plants in its eastern states. Together, the 15 countries of the European Union (EU) have held their carbon emissions to just 1 percent above the 1990 level in 1996, but industrial countries as a whole are likely to fall well short of their goal of limiting emissions to 1990 levels by the year 2000.[15]

The fastest growth in greenhouse gas emissions in recent years has occurred in the developing world, where per capita levels are low but industrialization is still gathering speed. (See Figure 4.) By 1996, emissions of carbon in developing countries were 44 percent over 1990 levels, and 71 percent over 1986 levels. Rapid economic growth, particularly in Asia and Latin America, is driving emissions up, as growing numbers of people are able to afford home appliances,

FIGURE 4

Carbon Emissions per Capita, Top 10 Emitting Nations, 1996

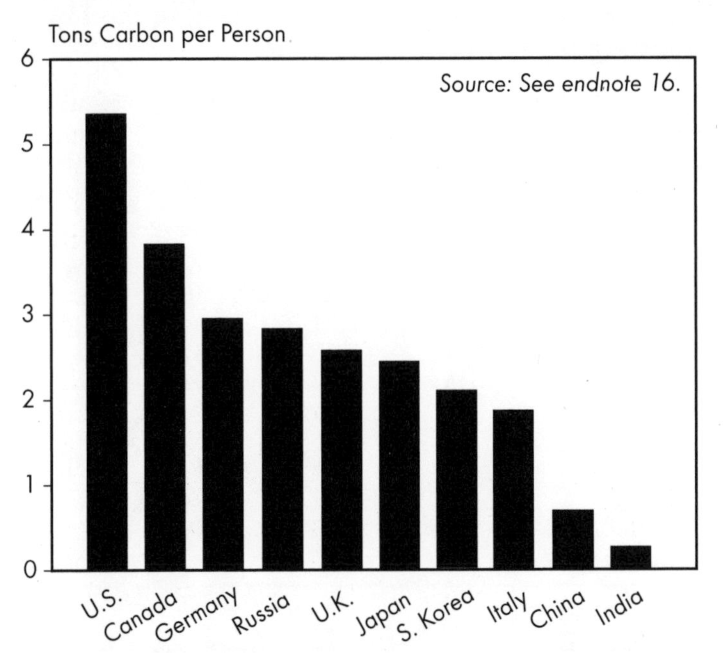

motorcycles, cars, and other energy-intensive amenities of a "modern" lifestyle. The emissions growth potential of developing countries remains substantial, however, as seen by the fact that China, despite its heavy dependence on coal, produces one eighth as much carbon dioxide per person as the United States, while per capita emissions in India are only about one tenth the level in Japan. (See Tables 1 and 2.) When these are added to industrial country emissions, the resulting global trend is sobering: the International Energy Agency (IEA) projects that, in the absence of any additional policy initiatives, worldwide carbon emissions from fossil fuels will exceed 1990 levels by 17 percent in the year 2000, and 40 percent by 2010, reaching 9 billion tons per year.[16]

Such trends and projections have led to angry charges and counter-charges between industrial and developing countries at recent climate negotiations, and have called into question the division of responsibilities agreed to when

TABLE 1

Carbon Emissions from Fossil Fuel Burning in the Top 12 Emitting Industrial Nations, 1996

Rank	Country	Total Emissions	Emissions per Person	Emissions per Dollar GNP	Emissions Growth 1990-96
		(million tons)	(tons)	(tons/million$)	(percent)
1	U.S.	1,433	5.35	198	8.8
2	Russia	414	2.81	627	-32.7
3	Japan	308	2.44	110	12.5
4	Germany	241	2.94	146	-7.6
5	U.K.	151	2.56	132	-2.0
6	Canada	115	3.82	180	5.3
7	Italy	107	1.86	94	0.6
8	Poland	95	2.46	452	-7.1
9	Ukraine	92	1.81	767	-55.8
10	France	91	1.55	74	-1.1
11	Australia	87	4.70	249	9.6
12	Spain	60	1.53	105	11.6

Source: See endnote 16.

the climate convention was originally negotiated. A resolution passed by the U.S. Senate in July 1997, for example, argues that growing emissions in developing countries will overwhelm any progress made on limiting emissions in industrial countries. It calls on the U.S. government not to agree to any additional limits in Kyoto unless the agreement "also mandates new specific scheduled commitments to limit or reduce greenhouse gas emissions for developing country parties." Representatives of developing nations reacted angrily to this suggestion at climate negotiations in Bonn in July 1997, noting that most industrial nations are missing the aims agreed to in Rio, and reminding their northern neighbors that their per capita emissions are still roughly 10 times those of developing countries.[17]

In the end, slowing global warming will only be possible if a cooperative North-South partnership supplants the finger-pointing. John Holdren of Harvard University likens the

TABLE 2

Carbon Emissions from Fossil Fuel Burning in the Top 12 Emitting Developing Nations, 1996

Rank	Country	Total Emissions	Emissions per Person	Emissions per Dollar GNP	Emissions Growth 1990-96
		(million tons)	(tons)	(tons/million $)	(percent)
1	China	846	0.68	234	29.3
2	India	250	0.26	184	37.6
3	South Korea	104	2.27	196	82.5
4	South Africa	95	2.24	n/a	16.8
5	Mexico	94	0.98	154	11.2
6	Iran	76	1.13	205	35.9
7	Brazil	65	0.41	75	25.1
8	Saudi Arabia	63	3.23	n/a	12.7
9	Indonesia	62	0.30	79	46.6
10	Kazakhstan	54	3.29	1,080	-33.8
11	Taiwan	48	2.23	n/a	48.2
12	Turkey	38	0.60	106	16.3

Source: See endnote 16.

world energy economy to a supertanker headed at full speed for a reef, asserting that "we all need to steer cooperatively, not argue who's at the wheel." Indeed, it is the industrial countries that are in the best position to pioneer a new generation of energy and transportation technologies that can be used to slow emissions growth. But developing countries need not lag far behind. And with the right policies, they could end up with cleaner, more efficient, and more economical energy systems than those currently in place in industrial nations. By investing soon in less-carbon-intensive technologies and infrastructure, developing countries can stay at the cutting edge of innovation, and avoid having to make two energy transitions in the space of just a few decades.[18]

Climbing Down

The roots of the global warming crisis lie in the energy revolution that transformed the world economy a century ago. Another energy revolution—one as dramatic as the changes that have swept the computer and telecommunications industries in the past decade—will almost certainly be needed to solve the problems created by the first one. Such fundamental changes in the world's energy system may seem farfetched, but are actually well within reach. Thanks to a powerful combination of government incentives and private investment, technologies such as synthetic materials, advanced electronics, and biotechnology are now flowing into the energy industry. The result is a host of new manufactured devices that efficiently and cleanly provide the energy needed to cook a meal, travel across town, or surf the Internet.[19]

One example of the kind of revolutionary technologies now on the horizon is a new kind of lightweight hybrid electric vehicle that uses small piston engines, turbine generators, or fuel cells as well as a highly efficient electric motor. Toyota has announced plans to put one such vehicle on the market in 1998—with a fuel economy of nearly 66 miles per gallon, or more than twice the current U.S. average. According to studies by the Rocky Mountain Institute in the United States, more advanced designs could exceed 100 miles per gallon.[20]

Similar technologies could be used in millions of residential and commercial buildings to generate electricity and heat while producing just 10 to 20 percent of the emissions that flow from today's large, centralized power generators. Micro-turbines and fuel cells now on the market can turn up to 90 percent of the fuel they consume into usable heat and electricity—compared to the 34 percent average for the large power plants now in use. The key to the efficiency of these devices is their location: unlike conventional generators, micro-power plants located in buildings have a ready use for their waste heat for water and space heating—a process known as "cogeneration." The key to their economics is the ability to produce such devices by the millions. In 1997, the U.S. company Allied Signal announced plans to market one such generator—at about one third the cost (and one tenth the carbon emissions), per unit of electricity generated, of a typical coal-fired power plant.[21]

Also evolving quickly are highly energy-efficient and cost-effective building and appliance technologies, which are now on the market. These include more efficient heating and cooling systems, better insulation, and advanced lighting and appliances. "Superwindows" that include special reflective material and can reduce a window's heat loss by 75 percent, compact fluorescent bulbs (CFLs) that use 20-25 percent of the electricity consumed by standard incandescent lamps, and refrigerators requiring half the power of models currently on the market are among the recent developments that can be employed to cut carbon emissions while saving money.[22]

A new breed of clean, emissions-free power plants is ready to go as well. The global wind power industry, which now provides less than 1 percent of world electricity, has become a $2 billion-a-year business, and is expanding at a rate of 25 percent per year. (See Figure 5.) Two decades of innovation have yielded wind turbines with tough fiberglass blades and electronic controls, with a generating cost that is comparable to that of new coal-fired power plants—and still falling. Since 1990, thousands of wind turbines have been installed in a dozen European nations—providing 5 to 10 percent of the electricity in some regions. India and China, the world's most populous countries, have also begun installing extensive "wind farms" in the past few years. And the wind resource is huge. According to our estimates, wind power could provide 20 percent or more of the world's electricity by the middle of the next century, displacing at least one third of today's fossil-fuel-fired power plants.[23]

The second-fastest-growing energy source today is solar power. (See Figure 6.) The worldwide manufacture of solar photovoltaic (PV) cells surged by an estimated 25 percent in 1997, reaching more than 110 megawatts (MW)—compared to just 47 megawatts in 1990. The cost of solar cells has fallen from more than $70,000 per kilowatt in the 1970s to $4,000 today and is expected to drop to as low as $1,000 per kilowatt in the next decade. An estimated 400,000 homes, most of them in remote areas not connected to power lines, already have solar power. In Japan, major housing companies, responding to new government incentives, plan to build 70,000 residential homes with silicon roofing tiles that generate enough electricity to meet most of their power needs. Similar programs are being launched in the United States and several European countries. Studies suggest that if the rooftops of existing buildings were covered by solar cells, they could supply half to three quarters of the electricity needed each year.[24]

These are but a few examples of the dramatic changes in energy systems now possible. If combined with improved public transportation, more compact cities, and new industrial processes, these changes could dramatically reduce carbon emissions while strengthening the global economy and creating millions of jobs. Although some economists argue that it will be expensive to develop alternatives to fossil

FIGURE 5

World Wind Energy Generating Capacity, 1980–97

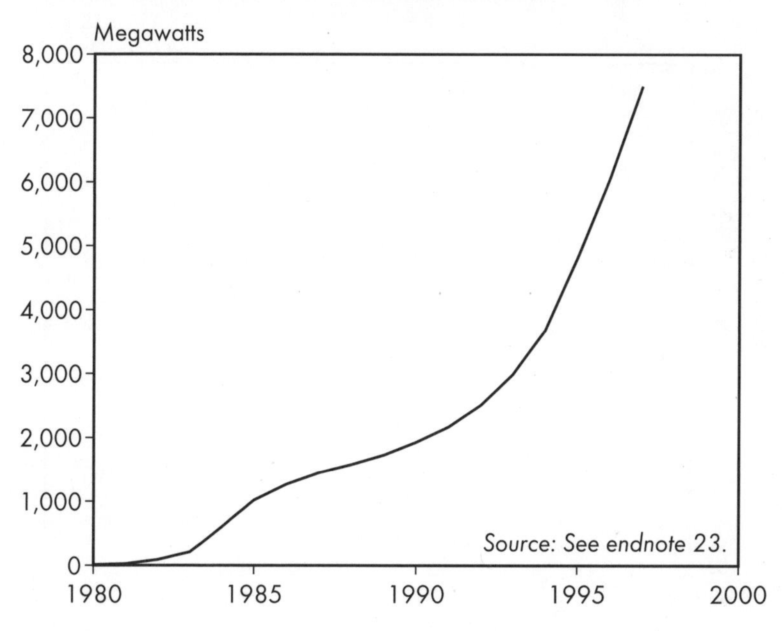

fuels—and that we should delay the transition as long as possible—their conclusions are based on a technological pessimism that is out of place in today's world. Just as automobiles eclipsed horses, and computers supplanted typewriters and slide rules, so can the advance of technology make today's energy systems look primitive and uneconomical. The first automobiles and computers were expensive and difficult to use, but soon became practical and affordable. The new energy technologies are now moving rapidly down the same engineering cost curves.[25]

Even in the face of such evidence, many business leaders continue to believe that any serious efforts to slow climate change will cause economic devastation. In this vein, Jerry Jasinowski, President of the U.S. National Association of Manufacturers, said at a September 1997 press conference that "this [climate] treaty could damage the U.S. economy more than anything we've seen for decades . . . it could raise

FIGURE 6

World Photovoltaic Shipments, 1971–97

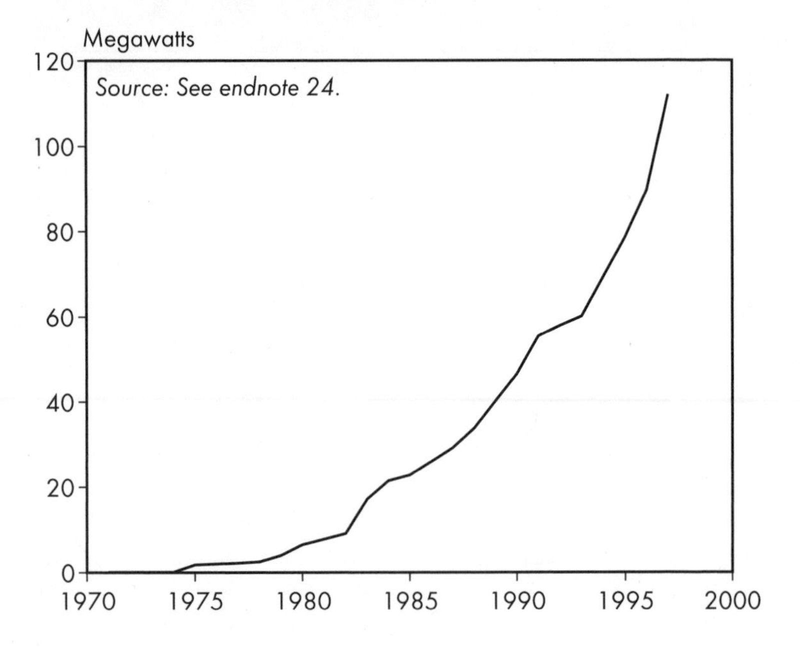

the cost of almost everything." That conclusion, however, is based on static "top-down" models of the economy that project a substantial cost, causing, for example, a 1 to 3 percent reduction in U.S. gross domestic product for a 30 percent reduction of emissions below 1990 levels by 2020. But such models are a poor reflection of the real world, which is far more dynamic. One shortcoming is their neglect of the potential for new policies to induce technological change, a process that has been described in the literature as "positive-feedback economics." This process was well illustrated by the phaseout of ozone-depleting substances during the early 1990s, when low-cost alternatives were found far more quickly and easily than economic projections had predicted.[26]

Another blind spot in many of these attempts to model the potential to reduce greenhouse gas emissions is that they assume a simple, solitary policy tool—putting a price tag on these emissions in the form of a tax on fossil fuels. This approach is used mainly because economists find it easier to model a tax than any other kind of policy. But the models tend to overstate the costs of such a policy, as they often assume the revenues will not be recycled to reduce other taxes. Moreover, in practice, it is far more effective—and efficient—to couple higher energy taxes with a broad range of other policies that remove the substantial existing barriers to new technologies, and provide incentives to get trends turned in a new direction. Nor do most models factor in the related benefits of reduced local and regional air pollution. According to economist Robert Repetto, adjusting the models to account for these factors suggests that well-designed efforts to deal with the climate problem could actually increase economic output.[27]

As economists retool their models, a flurry of more optimistic studies have appeared, which indicate that an appropriate mix of policies can deliver faster emission reductions at lower costs—avoiding the chaotic, costlier energy transformation that any further delay might eventually lead to. These studies demonstrate the potential for industrial countries to begin this process immediately—building on the roughly 30 percent gains in energy efficiency that most have made over the past two decades, and gradually bringing new technologies into use. For example, the U.S. Department of Energy found in a 1997 study that it would indeed be economically feasible to return U.S. carbon emissions to the 1990 level by the year 2010 (which would require close to a 10 percent reduction from 1997 levels) and that "a next generation of energy-efficient and low-carbon technologies" could spur aggressive carbon emissions reductions over the next quarter century.[28]

One major oil company has suggested a more ambitious sequence of events. In a 1996 report, the Group Planning unit of Royal Dutch Shell envisioned a "Sustained Growth" scenario in which fossil fuels dominate energy markets in the short run but renewable energy technologies "steadily progress along their learning curves, first capturing niche markets and, by 2020 become fully competitive with conventional energy sources." As the Shell analysts note, future trends do not always follow past patterns, and growing concern about climate change could cause energy markets to shift suddenly in a new direction.[29]

Even more bullish is a study of the European Union, conducted by the World Wide Fund for Nature in 1996, which concluded that a mix of new energy policies that promote energy efficiency and renewable energy could actually cut carbon emissions 14 percent below the 1990 level in 2005, without harming the economy. This compares with a 7 percent increase under a business-as-usual scenario. With this policy mix, the largest reductions in emissions derive from improved energy efficiency in manufacturing and increased reliance on cogeneration and renewable energy. A similar study, called *Energy Innovations*, published by a group of U.S. environmental organizations in 1997, found that U.S. carbon emissions could be cut to 10 percent below the 1990 level by 2010, while actually strengthening the economy (reducing annual national energy costs by $530 per household and creating nearly 800,000 jobs), compared to a 21 percent emissions *increase* on a business-as-usual path. Among the advances that could make such reductions possible are new gas turbines for power generation and the large-scale conversion of agricultural and forestry wastes to fuels and electricity. Government and independent studies in Canada, Japan, and Australia likewise point to the merits of a diversified policy mix for simultaneously reducing greenhouse gas emissions and improving their economies.[30]

These studies suggest that when the pivotal role of policy-induced technical change is given the attention it deserves, the opportunities to respond to climate change become greater than projected in most models—reaffirming the IPCC's conclusions that major emissions reductions are both technically and economically feasible. According to the panel, most countries still have ample "no-regrets" policy options: those that would be justified even without considering the need for climate protection. In many parts of the world, for example, efficiency improvements—and commensurate emissions cuts—of 10 to 30 percent can be achieved at little or no cost, while improvements of 50 to 60 percent are technically within reach. But the panel also warns that the risks of climate change, and the precautionary principle enshrined in the climate treaty, call for going well beyond "no regrets."[31]

The test now facing climate policymakers is to find the best combination of policies that will tap this quickly expanding potential for reducing greenhouse gas emissions. During the past decade, governments around the world have experimented with an array of new energy policies and other measures that can do so. Several of these policies have proved to be remarkably successful—and most are justified on economic grounds alone. Still, many look elegant on paper but fail miserably in practice. Others seem brilliant from an economic viewpoint but are political non-starters. And many policies work well in one nation or culture but may founder in another socioeconomic setting. Altogether, this early experimentation has provided a wealth of information that will guide national governments on the steep path from Kyoto.

Getting the Prices Right

In building a fossil-fuel-based energy system over the past century, most nations have developed a complex web of policies aimed at accelerating the extraction and use of coal, oil, and natural gas. Special subsidies and tax breaks for fossil fuels, as well as for road and car use, encourage heavier use of these fuels than would otherwise occur. Direct government financial support for conventional energy sources and technologies is currently estimated at $200 billion annually worldwide—more than half the value of all the crude oil produced each year. At a time when the goal should be to lower fossil fuel use, not increase it, these subsidies need to be reevaluated. According to the IPCC, industrial countries that cut subsidies soon could reduce their projected carbon emissions in 2050 by 18 percent while increasing overall income levels in the long run.[32]

The most dramatic fossil fuel price reforms so far have taken place in former eastern bloc countries, which under communist rule had the highest subsidies and consequently missed out on the conservation incentives provided by rising oil prices in the 1970s and 1980s. After their revolutions, however, these nations started phasing out fossil fuel subsidies. Poland has cut its energy supports by $3 billion annually, which contributed to a decline in coal use of more than 30 percent between 1987 and 1994. Greater cuts are possible, though rising natural gas prices and the pressure of mining lobbies have discouraged further switching, and coal use is now growing again. Russia, meanwhile, has slashed fossil fuel subsidies by more than half since 1990, and in the six years since, its carbon emissions have dropped more than 30 percent. Consumer energy prices are still heavily subsidized, though, and as a result, many apartment buildings still lack the most basic insulation and heating controls. However, renewed efforts to restructure the country's huge gas and electricity monopolies in 1997 may result in further reductions in these supports.[33]

Energy price reforms have been uneven in Western Europe, where fossil fuel subsidies still stand at more than $10 billion a year. The United Kingdom has implemented dramatic cuts, having overturned a law requiring the state-owned power system to purchase fixed amounts of coal from the country's producers; this led to a reduction in fuel supports from more than $7 billion in 1989 to virtually nil in 1995. Today the power industry is largely in private hands, and coal use has fallen 31 percent, replaced largely by North Sea natural gas, whose use has risen 62 percent. The country's carbon emissions fell steadily from 1990 through 1995, largely due to coal subsidy removal.[34]

In Germany, on the other hand, electric utilities are still required to buy the country's expensive domestic coal.

German coal subsidies have actually increased by more than half over the last 15 years, reaching nearly $7 billion in 1995. By the mid-1990s, these enormous sums, long justified by the objective of protecting coal-mining jobs, were costing roughly $70,000 per job—well above the salaries paid to German miners. The coal supports were paid by consumers until 1995, when they were converted into a federal subsidy due to be gradually phased out over the next decade. However, coal imports are growing and the planned reduction has been stalled by political opposition: subsidies are projected to reach a new high in 1998.[35]

Smaller but still important fossil fuel subsidies exist in other industrial countries, and are mainly aimed at producers. French coal supports, now being phased out, stand at around $722 million. Japan is also reducing subsidies for coal, though they remain at an estimated $149 per ton. Australia, the world's leading coal exporter, continues to prop up numerous uneconomical coal-fired power projects with an elaborate system of price supports and has stepped up its "clean-coal" program—which promotes the use and export of technologies that reduce emissions of sulfur and particulates, but that have only a limited effect on carbon emissions. The United States has gradually reduced some of its supports for fossil fuels, but they remain substantial—as high as $18.3 billion a year, according to a study commissioned by the Earth Council—and include special tax breaks for the oil and gas industry. Canada, too, heavily supports its oil and gas producers, through $6 billion in tax incentives each year, as well as direct funding of major development projects such as Hibernia in the North Atlantic, one of the most expensive oil fields ever developed.[36]

The greatest obstacle to fossil-fuel-subsidy reform is political. The Italian government, for example, could save $2.2 billion annually and avoid 12.5 million tons in carbon emissions by reducing subsidies to electricity—yet has failed to do so out of concern about the reaction of consumers and industrialists to any increase in prices. Still, such subsidies are likely to dwindle in the coming years throughout the European Union, as EU rules push member countries to reduce these supports and open the remaining energy monopolies to competition.[37]

Most developing countries also heavily subsidize fossil fuels, partly to shield poor consumers from world prices—but with the perverse effect of supporting costly imported fuels such as kerosene at the expense of technologies that could harness domestic energy sources, such as solar power. But reforms are under way, most dramatically in China, where annual oil and coal supports have been steadily reduced since the 1980s, falling from $24 billion to $10 billion over the last five years alone. These cuts are estimated to have lowered the overall energy intensity of the Chinese economy by 30 percent while slowing carbon emissions growth by 20 percent, according to the World Bank. Greater gains are possible: the Bank calculates that by eliminating these subsidies, China could save more energy annually by 2020 than it used in 1990. Other developing countries have also taken bold steps to phase out fossil fuel subsidies. India, for example, cut its own from $4.2 billion in 1991 to $2.6 billion in 1996. (See Table 3.) In Latin America, the broad market reforms introduced in the 1990s have led to dramatic reductions in consumer energy subsidies, including a complete elimination of fossil fuel supports in Brazil. Further progress is likely as Latin America's mining, petroleum, natural gas, and electricity industries are privatized in coming years.[38]

Removing subsidies for building and using roads can also reduce carbon emissions. Since the 1950s, all industrial countries have to varying degrees shifted their transportation focus from moving people—via rail, bus, and other mass transit—to moving cars—by building and expanding roads and road-related services. While these supports have not been fully tabulated, there is little doubt as to their large role in encouraging dependence on cars. The United States is "king of the road," annually handing out $55 billion in direct subsidies and another $66 billion for parking and other road-related services. Its support of road transport out-

TABLE 3

Fossil Fuel Subsidies in Selected Developing Countries and Countries in Transition, 1990–96

Country	Subsidy Rate 1990-91	Subsidy Rate 1995-96	Total Subsidies 1995-96
	(percent of market price)		(percent GDP)
Egypt	49	40	3.4
Eastern Europe	42	23	3.2
China	42	20	2.4
Russia	48	25	1.5
India	25	19	1.1
Indonesia	26	21	0.9
Mexico	28	16	0.7
Thailand	9	9	0.4
South Africa	11	4	0.3
Brazil	23	0	0.0

Source: See endnote 38.

weighs that of public transit by at least seven to one. In Japan, supports for road use have been estimated at $16 billion, with up to another $50 billion for road-related services and parking. Because of higher fuel taxes, such subsidies are far lower in Europe and in some cases negative, ranging from a road subsidy of some $13 billion per year in Germany to an annual net tax on road users of about $5 billion in both France and the Netherlands.[39]

Even countries with better records of prioritizing public transport are experiencing difficulties in fending off rising car use. Denmark has a relatively well-developed strategy, emphasizing investments in bikes, buses, and train infrastructure, but is investing even more in road expansion. Public transport is also losing ground to the car in the Netherlands, which has introduced policies that promote rail lines, bike paths, and other alternatives to the automobile but has still seen motor vehicle carbon emissions rise 15 percent since 1990. Japan actually includes road building in its national climate plan, even though such efforts will increase, not lessen, carbon emissions.[40]

One key to reining in road transport emissions is more integrated decisionmaking in transportation and land-use planning, an approach the United States, France, and the United Kingdom now mandate. Another is to combine public transit improvements with charges for road use. Governments are beginning to explore road pricing and other restrictions: examples of local initiatives to restrain car use can be seen in a handful of European and North American cities, such as Copenhagen, Denmark, and Portland, Oregon. Other successful efforts are found in former eastern bloc and developing countries, where cities such as Krakow, Poland, Curitiba, Brazil, and Singapore are encouraging the use of public transport via better planning, strategic investments, and congestion pricing.[41]

Taking this process of energy price reform a step further, several governments have levied taxes on carbon emissions and fossil fuels to better reflect their effects on the climate. According to a study supported by the Organisation for Economic Co-operation and Development, simply adjusting existing energy taxes in proportion to the carbon content of the fuels used could cut emissions by 12 percent. The IPCC estimates that higher gasoline taxes could cut most countries' carbon output by up to 25 percent by 2020, and as much as 60 percent where prices are currently low. Revenues from these fees can be returned to taxpayers through cuts in existing taxes, thereby increasing a nation's overall income.[42]

During the 1990s, five European countries have adopted carbon or energy taxes that, though weakened by exemptions and deductions, are showing results. The Dutch tax, which exempts renewable energy, is estimated by the government to be cutting carbon emissions 2 percent annually. Sweden's levy has increased biomass use, mainly for cogeneration, by 71 percent (See Table 4). Revenue from the Danish tax supports industrial energy efficiency, and is believed to be reducing emissions significantly. All three taxes are recycled to reduce taxes on income and wages.[43]

Elsewhere, energy and carbon taxes have been stalled

TABLE 4

Carbon/Energy Taxes in Industrial Countries, 1997

Country	Date Introduced	Current Rate	Exemptions
		($/ton C)	
Denmark	1996	1.0-3.7 industry 2.1-24.3 consumers	Electricity use
Finland	1990	1.9	Industrial raw materials and overseas transport fuels
Netherlands	1992	1.2-1.6	Large-scale natural gas use and renewable energy
Norway	1991	4.6-15.3	Onshore natural gas use and fuels for fishing, air, and freight transport
Sweden	1991	6.5 industry 13.1 consumers	Electricity use and some biomass use

Source: See endnote 43.

by industry opposition and the resistance of consumers to higher energy prices. In the United States, a 1993 proposal to tax the energy content of fossil fuels (dubbed a "BTU tax") was strongly opposed by industries, and finally replaced with a tiny 4.3-cent increase in gasoline taxes. A proposed tax in Australia that would have provided funds for an agency to promote energy efficiency and renewable energy met the same fate in 1994. On several occasions over the last five years, the European Commission has considered proposals for an EU-wide energy and carbon tax, and though it has not been approved so far, interest in the concept is growing in many European governments that would like to use some of the revenues from such a tax to reduce their burdensome payroll taxes. A proposed Japanese carbon tax was rejected in 1991, but the government's Environment Agency renewed its call for such a levy in the summer of 1997. Should sufficient political support gather, either tax could become the first step to a broader international tax on carbon emissions—treating the industries of all countries equally.[44]

Carbon levies are in the meantime gaining a tenuous foothold in former eastern bloc and developing countries. In 1993, for example, Poland imposed a small fee of two cents per ton on emissions from industrial plants and municipal enterprises. The revenues are used to pay for investments in energy efficiency and tree planting. Environmental groups, however, argue that the tax is too low to encourage industrial efficiency improvements. Also noteworthy is Costa Rica, which has installed a 15 percent national fuel tax, with part of its funds devoted to reforestation efforts.[45]

The lion's share of current energy taxes is levied on gasoline. Taxes as high as 60 cents per liter ($2.50 per gallon)—more than $1,000 per ton of carbon—are found in a number of European countries. Some, including Germany, Italy, and the Netherlands, have recently raised their fuel taxes. Others have had less success: the United Kingdom's fuel levy increase was removed in August 1997 by the new British government, and the United States, whose gasoline taxes are among the world's lowest at an average of 36 cents per gallon, has been unable to increase them. (See Figure 7.)[46]

Just as carbon tax funds are sometimes channeled to clean energy, gasoline tax revenues can be steered to public transportation—rather than road-related services, as they currently are. One nation doing this is Austria, where the fuel tax is partly earmarked for local rail systems. If more widely used, this policy could help industrial countries tackle their most problematic—and least addressed—source of carbon emissions.[47]

As these examples show, fiscal incentives are often most effective when they are combined. The impact of carbon taxes will be limited if fossil fuel subsidies remain in place; conversely, carbon taxes may not need to be raised so high if these supports are removed. Raising gasoline prices, similarly, will not stem car use sufficiently without

FIGURE 7

Gasoline Prices, United States, Germany, and Japan, 1978–96

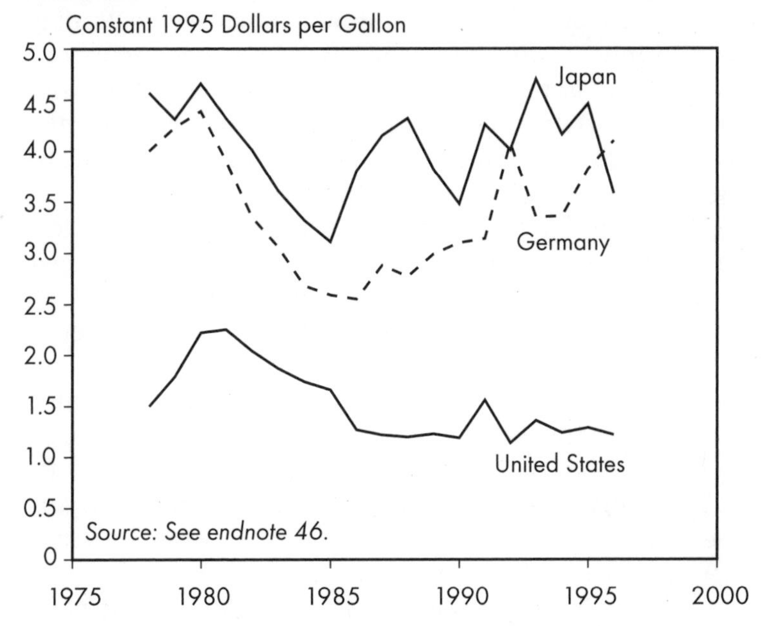

accompanying efforts to boost public transport and discourage road use. Experience so far suggests that comprehensive reform of government fiscal policies could spur rapid reductions in emissions, as well as strengthen national economies. Moreover, it can set the stage for a range of other climate policies to do their work.

Boosting Efficiency Standards

Another cause of rising carbon emissions is the tendency of purchasers of energy-using equipment to focus on initial prices without considering the lifetime fuel bills, which are often substantial. One widely accepted and time-tested means of overcoming this market failure is to set minimum efficiency standards that manufacturers must follow. If supported by higher energy prices and combined with incentives that motivate manufacturers and inform consumers, such standards can be steadily ratcheted up, leading to continuous efficiency improvements and emissions cuts—as well as long-term savings for consumers. The sooner they are strengthened, the greater their impact on the large number of cars, buildings, and appliances due to be manufactured and installed in the next decade.[48]

The need for efficiency standards is especially clear with automobiles, which dominate transportation's 21 percent share of energy-related carbon emissions. (See Figure 8.) Standards established in the 1970s improved efficiency markedly during the 1980s, but governments have not strengthened them since then—leading to a decline in fuel economy in the 1990s. In the absence of new standards, emissions from road transport are projected to double by 2020, with much of the increase occurring in developing countries. With higher car fuel economy, however, the emissions forecast could be cut by as much as a quarter.[49]

The only binding auto efficiency standards in place among industrial countries today are in the United States, where they have become nearly meaningless. (See Table 5.) Enacted in 1978 and gradually raised to 27.5 miles per gallon (8.6 liters per 100 kilometers) in 1985, these standards nearly doubled the fuel economy of new U.S. cars between 1974 and 1988 but have since remained essentially flat. Meanwhile, the average fuel economy of new vehicles has actually fallen, due partly to the booming popularity of gas-guzzling "sport-utility vehicles."[50]

Because of industry resistance to binding standards, the only fuel economy targets enacted in recent years have been weak voluntary goals. Twelve countries that have done so call for, on average, just a 10 percent efficiency improvement over 10 years, requiring little more than what manufacturers are already planning. To encourage customers to buy efficient cars, Austria and the Canadian province of Ontario have enacted "feebates," which either offer car buy-

FIGURE 8

World Carbon Emissions from Energy Use, by Sector, 1990

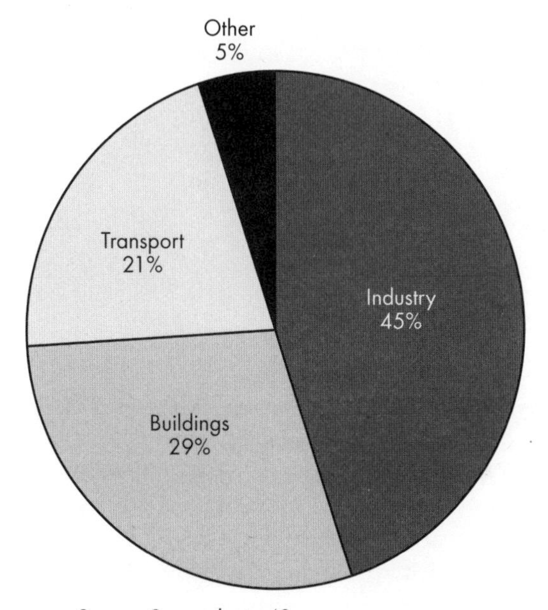

Source: See endnote 49.

ers a rebate or require them to pay a fee according to the car's fuel economy, but these are not large enough to have a substantial impact. The European Commission has proposed establishing an ambitious target of 5 liters per 100 kilometers (47 miles per gallon) for gasoline-fueled cars and 4.5 liters per 100 kilometers (52 miles per gallon) for diesel cars by the year 2005, compared to the current average of 8 liters per 100 kilometers (29 miles per gallon) for gasoline-powered cars, but has not been approved so far. In the developing world, only South Korea has mandatory auto efficiency standards.[51]

Buildings and appliances, which are indirectly responsible for 29 percent of global carbon emissions from energy use, present another large opportunity for cuts. Without fur-

TABLE 5

Automobile Fuel Efficiency Targets, Selected Countries, 1997

Country	Current avg.	Target			Timeline
	(L/100km)	(L/100km)	(mpg)	(percent improvement)	
Australia	11.0	8.2	28.7	–	2000
Canada	n/a	8.6	27.5	–	1985
France	8.0	5.8	39.2	–	2005
Germany	9.2	–	–	25	2005
Japan	11.9	7.4	31.7	–	2000
U.K.	9.0	–	–	10	2010
U.S.	8.6	8.6	27.5	–	1985

Source: See endnote 50.

ther measures, global emissions from these sectors are projected to as much as triple by 2050, mainly due to increases in floor space and the rising use of home appliances and office equipment—particularly in former eastern bloc and developing countries, where the majority of new buildings and appliances will be constructed and produced. The IPCC estimates that efficiency standards for buildings and equipment could hold their 2050 emissions to double the current level, while paying consumers back in full for the extra purchase cost within five years or less. Further reductions are possible if these "sticks" are reinforced with "carrots," such as tax incentives for retrofitting old buildings and purchasing more efficient equipment.[52]

Most industrial countries have building codes in place and have recently strengthened them, in some cases supplementing them with incentives. But the toughest demand a fraction of what is economically justifiable. Sweden's codes, enacted in 1983, remain the most stringent and have served as a model for other countries, but could be significantly tightened to keep up with the latest building technologies. New Dutch codes tap only a quarter of the achievable savings. Germany and Japan offer subsidies for buildings that meet or exceed standards, but their new codes are relatively weak.[53]

Other countries have struggled to implement still looser standards: a new code in Canada, which if widely used would cut building energy use by one fifth, has been adopted by just one province. In the United States, building codes are a state responsibility, but most are out of date, and many report a compliance rate of just 50 percent. This has lessened the impact of supporting efforts, such as a federal incentive program to construct buildings that are 30 percent more efficient than the code, and a program to help homeowners finance the upfront costs of retrofitting buildings by allowing them to include efficiency improvements in their home loans.[54]

Efforts to establish building codes elsewhere have been limited, but are accelerating. Russia has enacted, but is failing to implement, rules for efficient buildings. Mandatory building codes, most of which are modeled on the U.S. code, can also be found in the developing world in places such as Singapore, which reports high compliance, and China, which has also introduced tax incentives for the construction of energy-efficient buildings.[55]

While implementation is the largest challenge for building codes, adopting new standards and updating existing ones is the next step for appliances—which are a rapidly growing contributor to greenhouse gas emissions. A small though growing number of industrial countries have established appliance efficiency rules, but even the best performers remain well below what is economically feasible. The United States has since 1978 set criteria for more than 20 appliances, which have brought significant gains—including a tripling of refrigerator efficiency—and are projected to yield some $56 billion in energy cost savings between 1990 and 2015. Ambitious new U.S. refrigerator and air conditioner standards were enacted in 1997. These are supplemented by a "golden carrot" program that creates an early market for highly efficient products by holding competitions among manufacturers and guaranteeing purchases of the winning product. This approach is showing results both in the U.S. and Sweden, though overall sales have been lower than hoped. Recent European Union refrigerator standards, however, have been diluted to below half the proposed levels because of industry opposition.[56]

Even greater opportunities for appliance efficiency gains can be found in developing and transitional countries, where sales of these products are skyrocketing, and the equipment on the market today is often shoddy. Refrigerator standards in eastern Europe are gradually being harmonized with those in western Europe, but are still being widely flouted. Several developing countries have set appliance standards, including Mexico, whose air-conditioning rules are based on previous U.S. standards. In the Philippines, where mandatory labeling and standards were introduced in 1993, air-conditioning standards have been strengthened and expanded to include imported products, and are now being used as a model for other developing countries.[57]

Office equipment—the fastest-growing source of emissions from commercial buildings—lacks standards entirely, although a U.S. computer-labeling program to encourage efficiency now covers 60 percent of all personal computers. Voluntary office equipment targets have been set in Switzerland and Japan—whose office electricity consumption is believed to be one of the country's biggest carbon emissions problems. The Tokyo-based Global Environment Information Center, under the auspices of the United Nations University and Japan's Environment Agency, has recommended that Japan take the lead in calling for the establishment of international standards for office equipment and other appliances.[58]

Experience shows that even in the current era of market-oriented policy, efficiency standards are essential—particularly in developing countries where the temptation to continue purchasing cheap, inefficient equipment is often irresistible. Unfortunately, however, progress in setting new efficiency standards over the past five years has been limited, with some headway in buildings, and less in appliances. The fastest-growing energy users—automobiles—have had the fewest encounters with standard-setting.

Forging Industry Covenants

Industry, which accounts for 45 percent of the carbon emitted from energy use, incorporates thousands of distinct devices and processes and demands a more flexible approach to cutting emissions than prescriptive standards can provide. It would be more effective in this case for governments to set targets that companies themselves then decide how to meet. If sufficiently stringent, covenants between government and industry can tap a carbon savings potential that, despite two decades of efficiency improvements, remains great. The IPCC estimates that industrial countries could reduce carbon emissions from industry 25 percent below 1990 levels in the short run, and considerably more in the longer term, by upgrading manufacturing facilities with the most energy-efficient technologies available. If performed in conjunction with the normal turnover of equipment, these upgrades would entail minimal costs. The potential is even greater in developing countries and those in transition, whose industries are generally less efficient than those of industrial countries.[59]

Over the past five years, governments and industries have signed some 200 such agreements aimed at reducing industrial greenhouse emissions. These vary widely in scope, ambitiousness, reporting requirements, and enforceability. The more successful ones resemble formal contracts between government and industry, with specific goals, required reporting, and penalties for noncompliance. Those purely voluntary in nature have been much less effective.[60]

The most ambitious industry covenants to date are found in the Netherlands, where there is a long tradition of such agreements; since 1989 the government has reached 28 individual "Long-Term Agreements" with more than 1,000 companies covering 90 percent of the country's industrial energy use. Calling for, on average, 20 percent efficiency improvements between 1989 and 2000, the targets are negotiated with each industry group—ranging from major energy users, such as chemicals and paper, to smaller enterprises, such as laundries and coffee roasters. (See Table 6.) The Dutch covenants allow for extensive inventories conducted by a government agency, which receives annual reports from each industry and monitors their progress. The agreements are considered contracts under civil law, which allows for the adoption of mandatory regulations if the targets are not met. According to a 1995 government study, two thirds of the agreements have led to an average 9 percent efficiency gain from 1989 levels, and are on schedule to meet the year 2000 target. The energy savings achieved have been greater than expected, and the government is now considering even tougher goals.[61]

Germany's voluntary programs, which include 19 industry sectors—totaling 80 percent of industrial energy use—also have specific targets and are independently monitored. The government projects that the programs will reduce carbon emissions by 50 million tons by 2005, but independent evaluations criticize their lack of regulatory backup and question whether they will stimulate improvements beyond "business as usual." Tighter targets are being discussed, and the government now says it will enact mandatory measures if the current goals are missed.[62]

The United States is having even less success with its voluntary agreements, which consist of general commitments, optional reporting, and no government authority to decree mandatory measures. Its "Climate Wise" industry program has attracted companies representing a mere 7 percent of industrial energy use, few of which have developed action plans. The U.S. government's "Climate Challenge" program with electric utilities has resulted in a handful of innovative programs to promote renewable energy and carbon sequestration, but suffers from declining participation

TABLE 6

Industry Covenants, Selected Industrial Countries, 1997

Country	Share of Industrial Energy Use Covered	Targets	Reporting Requirements
	(percent)	(average)	
Netherlands	90	20% efficiency improvement from 1989 level by 2000	Govt. agency, required
Germany	80	20% emissions reduction from 1990 level by 2005	Independent institute, required
Japan	60	10–20% emissions reduction from 1990 level by 2010	Govt. agency, required
United States	7	None	Industry self-monitoring, not required
Canada	70	None	Industry self-monitoring, not required
Australia	30	None	Industry self-monitoring, not required

Source: See endnote 61.

and is achieving a fraction of the carbon savings originally projected.[63]

More successful is the U.S. "Green Lights" lighting efficiency program, which since 1989 has committed participants to upgrade the lighting on at least 90 percent of their floor space for which such improvements are cost-effective. The program now has 2,300 participants covering one in every 14 buildings, averages lighting energy savings of 48 percent, and in 1996 is estimated to have removed the annual equivalent of the carbon emissions of 400,000 cars. Government funding cuts have caused recruitment to level off, however, and the reported savings are below what was originally projected. Still, the program's structure is being replicated abroad, with support from the U.S. Environmental Protection Agency. In the space of two years, Poland's Green Lights program has increased CFL sales by almost 50 percent, saved $79 per household in utility bills, and cut more than 130,000 tons in carbon emissions. China's aggressive Green Lights program includes plans to institute efficiency standards: the country could potentially save 40 percent of its projected lighting energy use through equipment upgrades.[64]

Also weak are the industry covenants of two other carbon-intensive economies: Australia and Canada. These programs are characterized by vague commitments, optional monitoring and enforcement, and limited public access to information. Both programs have been severely criticized by independent reviews for their passive governmental role and an apparent lack of serious commitment by industry participants. Most companies in the Canadian program have only submitted letters of intent, with just a few detailing extensive action plans. The Australian program, meanwhile, assumes industries will achieve no business-as-usual efficiency improvements, thereby inflating the projected results.[65]

Somewhere between those of the Netherlands and Australia lie Japan's industry agreements, which cover about 60 percent of manufacturing and specify targets and actions in 15 sectors. Their overall goal is a 10 percent efficiency improvement by 2000. These pacts were strengthened in 1996 when Keidanren, Japan's industry association, announced a voluntary environmental plan that included energy efficiency targets for 37 industrial sectors, and again in 1997, when it announced that its 137 members would aim to reduce emissions by 10 to 20 percent below 1990 levels by 2010. This came shortly after the government's formal recommendation that industry improve its efficiency 1 percent annually, with companies unable to do so required to submit a report explaining their failure. These efforts are supported by a two-decade tradition of tax credits and low-interest loans for industries installing energy-efficient equipment. The country's history of industrial efficiency improvements and the seriousness with which industry has regarded previous agreements with government suggest this approach may yield positive results in Japan.[66]

At least eight developing countries have voluntary efficiency programs under way for industry as well as for buildings and appliances. China has been especially innovative in encouraging industrial efficiency by basing worker bonuses on achieving standards in several sectors, monitoring energy use through a network of energy conservation centers, and setting quotas on energy use. These measures have stimulated nearly $6 billion in energy efficiency investments.[67]

The above experiences reinforce what the IPCC has concluded: that industry agreements are only the first line of defense in improving industrial energy efficiency and by no means lessen the need for additional measures. To be effective, moreover, they must include specific and ambitious goals, required reporting, independent monitoring, and penalties for noncompliance—most of which are still rare. And they will need to support other policies, not replace them, as many of the current programs were designed to do.[68]

Supporting New Energy Supplies

Avoiding dangerous climate change will depend in large part on our ability to develop new energy supply systems quickly. New technologies, such as cogeneration, are now becoming available that provide electricity far more efficiently and cut carbon emissions by between 60 and 80 percent. In addition, sunlight, wind, and other renewable resources could meet most of the world's energy needs by the middle of the next century. The IPCC estimates that the aggressive development of economically competitive alternatives to fossil fuels could lower by two thirds the cost of a 20 percent carbon emissions reduction. While appropriately pricing fossil fuels will help pull these new technologies into the market, a push is also needed to overcome the many barriers currently impeding the adoption of new energy systems.[69]

The first step is lowering the cost of the new technologies. Already, major technological gains and cost reductions have resulted from two decades of publicly funded research and development into renewable energy technologies that rely on sunlight, wind, and biomass. Though such funding has recently risen, it stands at around $900 million—less than half the 1980 high, and under 9 percent of total energy R&D budgets in industrial countries. (See Figure 9.) (By comparison, in the computer industry, Microsoft alone spent $860 million on research and development in 1995.)[70]

Even more important than increasing support for R&D, however, is surmounting the obstacles to deploying renewable energy technologies. The most formidable of these are the high initial cost of installing the equipment and the refusal of many electric utility monopolies to allow them to sell their power. Fortunately, the 1990s have given birth to a new generation of renewable energy policies that appear to be overcoming these obstacles—with the initiative shifting from the United States to Europe and Asia.[71] (See Table 7.)

Two policies in particular are proving their ability to bring renewables to market quickly. Tax incentives—subsidies to partly cover initial capital costs—prime the pump for these newer technologies, and can be phased out once they gain a foothold. Access to power markets at fixed or special prices is a second mechanism that has helped the new technologies off the ground in several countries. These two tools—alone or, more dynamically, in tandem—can have a demonstrable effect on emissions while creating considerably more jobs and exports than the conventional technologies they replace.[72]

Tax incentives have been widely used to promote renewables since the 1970s. Consisting mainly of tax credits, deductions, and exemptions for investment and generation, they can be a useful way to compensate for the high initial capital costs of new generating technologies such as wind power or geothermal energy, and can lure new compa-

FIGURE 9

Renewable Energy R&D, IEA Countries, 1974–95

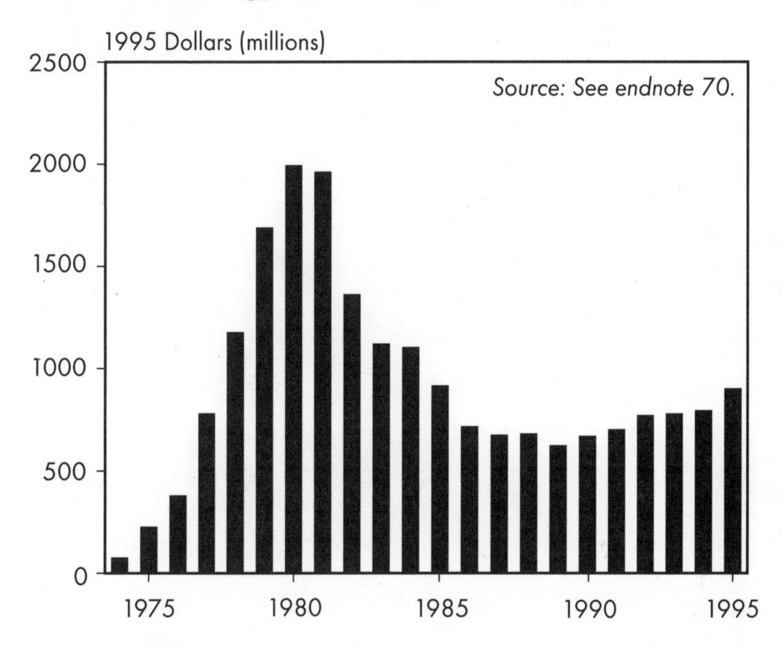

nies into the business. When used in a careful and consistent manner, tax incentives have proven helpful in bringing a technology to maturity.[73]

The effective use of tax incentives to stimulate the use of new energy technologies is best seen in Denmark, which in 1979 began to offer wind generators a 30 percent tax deduction for installing new wind turbines. These incentives, which were phased out as wind power costs declined, led to the installation of 2,500 wind turbines by 1992. New incentives for wind power were established in 1992, including a subsidy of 4.2 cents per kilowatt-hour of electricity generated and rebates on the national energy, value-added, and carbon taxes. These measures allowed Denmark to obtain 6 percent of its electricity from wind power by 1997, and have helped it become the world's leading manufacturer and exporter of wind turbines. Similar subsidies for cogen-

TABLE 7

Policies to Promote Renewable Energy, Selected Industrial Countries, 1997

Country	Tax Incentives	Market Guarantees
Australia	None	10% premium from some utilities
Austria	Generation credit for PV	Premiums in some regions
Canada	Accelerated investment tax depreciation	Premiums in some provinces
Denmark	15% investment credit for wind $.042/kWh generation credit for wind Energy, CO_2, value-added tax rebates	85% of retail electricity price for wind
France	Tax credits for PV, wind	$.048/kWh premium for wind
Germany	50–67% investment credit for PV (@$.10/kWh)	65–90% of retail electricity price for PV, wind
Japan	50–67% investment credit for PV	100% of retail electricity price for PV, wind (@$.18/kWh)
Netherlands	$.08/kWh generation credit for wind 11.5% energy tax deduction; CO_2 tax exemption Tax exemption for green-fund investors	None
Sweden	35% investment credit for wind; $.012/kWh generation credit	None
United Kingdom	None	$.063–.084/kWh premium
United States	$.015/kWh generation credit for wind, biomass $.014/kWh for state and municipal wind, "closed-loop" biomass 10% investment credit, for PV, geothermal Federal and state accelerated tax depreciation	None

Source: See endnote 71.

eration—up to half the cost of the equipment is deductible—have given it a 40 percent share of electricity use, and the government is aiming for a 65 percent share by 2005. An independent review of the Danish climate plan projects that by 2010 all of the country's power not provided by wind energy may come from cogeneration.[74]

In the United States, renewable energy tax credits have had a more erratic history. In California in the early 1980s, renewable power generators received a 10 percent federal investment tax credit, a 15 percent business energy investment tax credit, a five-year accelerated depreciation for wind systems, and a 25 percent state energy investment tax credit. In response to this policy and accompanying high purchase prices, some 12,000 megawatts of renewable capacity were installed during the 1980s, enough to supply electricity for some 10 million homes. The abrupt cutback of federal incentives late in the decade caused the market to collapse, and sent several companies into bankruptcy. Nevertheless, while the tax incentives would have been more effective if they had been more modest to begin with and reduced more gradually, they did help launch renewable energy in the United States, and attracted the interest of many developers in other countries. Today, power from solar, wind, biomass, and geothermal energy provide more than 9 percent of California's electricity.[75]

More recently, the U.S. Energy Policy Act of 1992 extended the 10 percent federal income tax credit for producers of solar and geothermal power, and offered operators of power projects based on energy crops and wind power a federal tax credit of 1.5 cents per kilowatt-hour generated during the first 10 years of the project. This subsidizing electric generation rather than capital investment creates more incentive for operators to lower their costs. Similarly, city- and state-owned utilities, which are exempt from federal income taxes, receive a federal payment of 1.4 cents per kilowatt-hour for wind and "closed-loop" biomass projects under the new law. However, a period of low fuel prices, overcapacity in power markets, and the uncertainties of electric industry restructuring has slowed development, and very few renewable power projects have been able to take advantage of the new incentives. U.S. wind power capacity has actually declined during the 1990s, as some older turbines have been dismantled.[76]

Tax incentives in the Netherlands have already boosted cogeneration to 30 percent of total energy use and are now being employed in a push for renewables. From 1986 to 1995, the Dutch government provided capital subsidies of up to 35 percent for new projects. The grant was replaced in 1996 by a guaranteed price of 8 cents per kilowatt-hour for wind projects of 2 megawatts or less. It adds an environmental tax credit of 2.7 cents per kilowatt-hour and a fossil fuel tax refund of 1.5 cents to the market price of 3.9 cents, effectively doubling the price paid to wind generators. In addition, developers are given an accelerated depreciation allowance for the wind turbines and can deduct the customary value-added tax from their remaining tax payments. The Dutch government also offers income tax exemptions for investors in approved green power projects and reimbursement for half the cost of wind resource assessments. The Dutch efforts have been plagued by frequent policy changes, however, and it remains to be seen whether a stable market for renewable energy can be achieved.[77]

While many countries are now shifting their tax incentives from investment to generation, Sweden offers both capital and generation subsidies for renewables. The former supports 35 percent of the capital cost of turbines above 60 kilowatts, and the latter pays $.012 per kilowatt-hour. Italy provides capital subsidies for renewable systems, but owing to budget shortfalls it has yet to disburse any funds. Canada allows accelerated depreciation of capital for renewables equipment, and is considering additional tax deductions. A number of countries are using tax incentives to promote building-integrated solar electric systems. Switzerland subsidizes the installation of such systems on school buildings as part of its strategy to have at least one solar electric system in each of its 3,000 villages by 2000.[78]

One of the most effective new policies to support new generating technologies in the 1990s is the so-called electricity "feed law," which sets a fixed price at which small renewable energy generators are provided access to the electricity grid. The most successful example of this policy so far is Germany's 1991 statute, which obligates utilities to purchase renewable electricity at up to 90 percent of the average purchasing price for end users. Guaranteeing a premium of 17 pfennigs (10 cents) per kilowatt-hour for wind generators, the act spurred the addition of roughly 2,000 MW of wind power capacity between 1991 and 1997, allowing Germany to surpass the United States as the leading wind power generator. In the first few years, tax incentives and other government programs helped get the new technologies off the ground, but today the feed law alone is enough to support a sizable market. The law's impact has prompted some utilities to fight to repeal it, but strong public support has kept it in place. The law has also reportedly created at least 10,000 new jobs in the wind power industry.[79]

Germany has created 10,000 new jobs in its wind industry since 1990.

Similar electricity feed statutes have begun to take hold in a half dozen other European countries and, according to German energy analyst Andreas Wagner, "have become the decisive criteria for renewable energy development in Europe." These laws vary widely in the technologies covered and the prices paid—and therefore in their effectiveness—but they have a proven ability to overcome the biggest single obstacle to the new technologies: the understandable tendency of monopoly electric companies to block them from the power market. Such codes have worked best when combined with tax incentives in the early years. One recent success story is Spanish policy, which provides a price of more than 7 cents per kilowatt-hour for wind, solar, biomass, and small-hydro electricity; it helped make the Spanish wind power market the fourth largest in the world in 1996.[80]

Another market access policy, the United Kingdom's Non-Fossil Fuel Obligation, requires utility companies that distribute power to purchase small but steadily growing portions of renewable power for their electricity portfolios. Under the program, which is supported by a 10 percent levy on fossil-fuel-based electricity, independent power suppliers compete for contracts to supply power from wind turbines and other renewable energy sources. These contracts guarantee a price above that paid to fossil fuel generators and last for a period of eight or more years while the price support steadily drops. Initially designed to prop up the nuclear industry, the program has encouraged a small rise in wind power installations of more than 87 MW between 1992 and 1996. But an additional 323 MW under contract have been held up because of local opposition to the siting of wind farms and financial constraints resulting from a highly competitive bidding process. The British government boasts of the substantial declines in the prices paid for renewable power in the course of four successive bids, but this may in part reflect technological improvements spurred by the much larger continental European market.[81]

Another approach to combining market guarantees with tax incentives can be seen in the ambitious solar rooftop program launched by the government of Japan in 1993. Its aim is to install 70,000 solar power systems on the rooftops and facades of buildings by 2000. Under this program, the government subsidizes half the cost to consumers, and two thirds of the cost to commercial building owners, of installing the solar systems. But it also requires electric utilities to buy electricity from the solar system owners at the full retail price of power. Approximately 3,500 such homes were built in 1994-96, most of them based on a new generation of residential homes with solar rooftop tiles that were developed just for this market.[82]

Japan has now become the world's leading PV user, driving the entire global market for solar cells up by an estimated 30 percent in 1997, with an estimated 9,400 homes subsidized that year. Encouraged by recent government studies estimating that solar and wind power could meet 55 to 65 percent of the country's electricity needs, Japan hopes to become a leader in developing carbon-free energy sources, just as it was in deploying air pollution control technologies in the 1970s. The government has made PV market dominance an official goal, and in August 1997 it announced plans to quintuple subsidies for solar and wind energy projects, although it is unclear what form they will take.[83]

The Japanese rooftop program has a number of rivals, including those in several European countries, such as Denmark, the Netherlands, and Switzerland. In Germany, a "1,000 roofs and facades program" provided a capital subsidy of 65 percent and a guaranteed power purchase price, spawning 5,000 systems between 1990 and 1995. Program funding has declined since then, however. More effective of late are the German Compensation Bank's 50,000-roof low-interest loan program and the adoption by 41 cities of rate-based incentive programs, which require utilities to buy surplus solar electricity at 90 percent of the average electricity price. Partly as a result, German PV use has increased nearly 10-fold since 1990.[84]

The United States, the world's leading manufacturer of solar cells, has so far had limited success in boosting domestic PV use. In June 1997 President Clinton proposed a "Million Roofs Program" to support tax incentives and loans for solar installations, but it remains to be seen whether they will be large enough to be effective. Fourteen states, meanwhile, have established "net-metering" laws similar to those of Germany, allowing consumers to sell surplus solar electricity back to their local utility company by in effect turning back their electric meters.[85]

In a number of industrial countries—including Australia, the Netherlands, Sweden, and the United States—the competitive restructuring of power markets has made it difficult to use regulatory tools to promote the new technologies, which must compete with low-priced power from highly efficient natural-gas-fueled plants. This has led to innovative experiments in "green power" pricing and marketing predicated on surveys suggesting considerable consumer interest in buying electricity from renewable energy sources. In green power *pricing*, monopoly utilities offer renewable energy to consumers at a slightly higher price. Utilities in several countries have begun pilot projects to test this concept. In green power *marketing*, competitive electricity suppliers are given the option of selling renewable power directly to consumers—at whatever price the market will bear. In both cases, public willingness to pay a little extra for these technologies at first has been stronger than predicted, and some early pilot programs have been oversubscribed. To ensure that these programs cause a real shift in the energy mix, efforts are under way to establish disclosure requirements and certification programs in Australia, Sweden, and the United States.[86]

Some developing countries have moved aggressively to increase their use of renewables and cogeneration. One could, in fact, argue that their embryonic energy systems position them well to lead the way in harnessing these new sources, perhaps even to leapfrog industrial nations. Brazil, for example, is home to what may be the world's most ambitious—and perhaps most controversial—renewable energy program. A 20-year-old price support program for the sugarcane-derived alcohol fuel ethanol has allowed it to displace half the gasoline used in the country's cars. Some 4 million cars run exclusively on ethanol, and the remaining automobile fuel is 22 percent ethanol and 78 percent gasoline. While lowering carbon emissions by 20 percent, the program also helps support Brazil's sugarcane industry. The subsidies have been criticized for the drain they impose on the government, but President Fernandez Henrique Cardoso announced at a June 1997 Special Session of the U.N. General Assembly in New York that he would continue to support the program through government procurement and tax breaks for some ethanol-fueled vehicles. One expert estimates that with this renewed effort alcohol-fueled cars could comprise 7 to 8 percent of the new cars produced in Brazil by 1998. Brazil is also promoting wind and solar power in its new energy plan.[87]

A package of new policies, dominated by generous tax breaks, has led to the explosive growth of India's domestic wind industry in the 1990s. The combination of an enticing investment tax credit, a special agency devoted to renewables that provides loans and grants, and a guaranteed purchase price pushed India's wind-generating capacity from 39 MW in 1992 to 820 MW in 1996, making the country the fourth-leading user of wind power. Additional growth is expected from 22 joint ventures now under way, some of which involve the world's largest wind turbine companies. Although the Indian program has been chastised for its overly generous subsidies, it still stands out for having created the first sizable wind industry in the developing world.[88]

India and China are moving aggressively to increase renewables and cogeneration use.

China, meanwhile, appears to be taking early steps along a path to leadership in renewables and cogeneration. Renewable energy (including biomass and hydropower) accounts for a full quarter of China's energy use, and nearly half the total in rural areas. The government has successfully promoted biogas and small hydropower for decades, but has just begun serious efforts to promote solar and wind energy. Several provinces—including wind-rich Inner Mongolia, Xinjiang, and Guangdong—offer favorable power purchase prices for windfarms, while low-interest loans are now available in some regions for off-grid renewable energy systems. By 1997, the use of wind and solar power was growing rapidly in several provinces, mainly as a result of projects being financed by foreign governments and international agencies. China is also promoting cogeneration, which as of 1994 provided 12 percent of its electricity. The Chinese government recently required that all large industrial boilers be converted to cogeneration, a policy that could serve as a model for the rest of the world.[89]

As these experiences indicate, a package of regulations and incentives supporting renewable energy and cogeneration can be an effective weapon in combating climate change—especially if complemented by other measures. Such synergy is lacking in Australia, Canada, and the United States, whose modest supports for these new energy sources are undermined by low energy prices. In contrast, integrated renewable energy and cogeneration policies in the United Kingdom, Germany, the Netherlands, and Denmark have been reinforced by fiscal reforms, allowing them to lead the world in commercializing the new technologies.[90]

Storing Carbon

The world's forests can also be enlisted in the climate change battle. Forests absorb carbon through photosynthesis and release it through respiration and decay, and can serve as carbon "sinks" if properly protected and managed. Deforestation, mainly in the tropics, currently adds about 2 billion tons of carbon—equal to one third of the total emitted from fossil fuel burning—to the atmosphere each year. These emissions are partly offset by temperate forests, which may be absorbing up to 1 billion tons of carbon.[91]

According to the IPCC, a concerted effort to protect intact forests, better manage commercial forests, increase the rate of reforestation, and raise the efficiency of wood-product use could convert forests back into net sinks for periods of between 50 to 100 years, offsetting as much as 15 percent of the carbon emitted between 1995 and 2050. The potential for sequestering carbon is greatest in tropical forests (80 percent), but is substantial in temperate (17 percent) and boreal forests (3 percent) as well. Studies suggest that this could represent a relatively inexpensive way to slow, though only temporarily, the buildup of carbon in the atmosphere—costing approximately $4-5 per ton of carbon.[92]

Because the scope for adding to the carbon pool simply

by managing existing forests is limited, sequestration strategies will need to emphasize an increase in forest area. But most industrial countries do not, or are only beginning to, integrate climate change concerns into their long-term forestry plans. Few acknowledge the need to aggressively plant trees, and fewer still have adopted policies that help them move in that direction. Indeed, most are struggling to maintain existing levels of forest cover.[93]

This general failure to see forests as a climate regulator is most evident in North America, where some degraded forest has been restored but little new forested land has been added. The United States, which has the most potential to sequester carbon in temperate regions, focuses on managing its existing forests rather than adding to forest area, offering only a small initiative to encourage tree planting on private lands. Funding for this program has been sharply reduced, and the nation's remaining old-growth forests—a small carbon sink—continue to be clearcut. Despite several private tree-planting incentive programs, by the year 2000 U.S. forests are projected to sequester 23 million tons of carbon less than originally expected, and forest loss is expected to continue into the next century. This implementation gap also appears in Canada, which according to an in-depth review has carried out none of its plans to enhance its carbon sinks. This failure, together with a rising incidence of forest fires and pests, has made Canada's extensive forests a net source of carbon.[94]

A similar lack of commitment to carbon storage plagues the industrial countries across the Pacific. In Australia, the "One Billion Trees Program" sets no requirements for increased forest cover. Land clearing there is estimated to contribute some 27 percent of total greenhouse gas emissions. Japan has set ambitious carbon sequestration goals but emphasizes the protection of existing forests in its current plan, has a limited afforestation plan, and is losing old-growth forest to development.[95]

The situation is slightly better in Europe. France has increased its forest area in recent years, and subsidies for reforestation are expected to store some 2 million tons of carbon by 2000. In Sweden, where growing forests offset the equivalent of 90 percent of the country's carbon emissions, a new policy requires that a number of forestry practices be altered to limit these emissions. The United Kingdom promotes afforestation through a number of incentive schemes, and has achieved a steadily increasing forest cover. Germany supports forest expansion partly through a "new afforestation bonus" of $773 per hectare per year; its forested land is increasing slightly and offsets some 4 percent of emissions.[96]

Stronger links between forest and climate policy have been made in two countries with little forest area: Denmark and the Netherlands. Denmark aims to double its forest area and provides a range of tree-planting subsidies to achieve this goal. The government estimates that over time this effort will offset around 5 percent of current carbon emissions. The Netherlands' forestry plan targets an increase in forest area of 20 to 25 percent by 2020, through a mix of government funds and voluntary actions—including a nascent system of carbon credits whereby companies can use tree planting to offset emissions requirements. Still, implementation of the program has reportedly been slow.[97]

The use of "carbon credits" to attract funds for sequestration has been avidly promoted in Costa Rica, where several forestry projects have been included in the United States' pilot "joint implementation" program, in which U.S. companies undertake carbon emissions reduction projects in other countries in the hope of eventually being able to use them to help meet domestic emissions goals. The cumulative effect of these efforts could be significant: one of the forest restoration efforts, for example, is expected to sequester 5 million tons of carbon—about five times Costa Rica's yearly manmade carbon emissions—over the project's lifetime.[98]

Several other countries in Latin America, Central and Eastern Europe, and Asia are also beginning to attract private capital for sequestration through "joint implementation" projects with the U.S. and other countries. While mostly focused on carbon storage to date, joint implementation also includes energy projects. Though currently limited in number and in their ability to attract larger investment flows, these projects may eventually serve as the basis for a global emissions trading scheme that is being discussed in the current protocol negotiations. Meanwhile, Costa Rica has established its own greenhouse gas fund, issuing "certified tradable offsets," which are already being traded on the Chicago Board of Trade, in exchange for investment in projects that sequester carbon. The country hopes to expand the fund into a worldwide carbon bank through which countries trade carbon emissions permits, and is cooperating with the World Bank and the government of Norway to this end.[99]

Costa Rica has established a fund for carbon sequestration projects.

Another way to finance carbon sequestration may be to meet a potentially large demand among industrial-country consumers for sustainable forestry projects. Three companies, two in the United Kingdom and one in France, have formed in response to surveys suggesting that a large number of people would pay a small premium on their energy bills to support tree-planting programs. These firms intend to undertake forestry projects in both industrial and developing countries, and plan to have them assessed independently by environmental experts. But other sources of revenue can be drawn upon too, as Poland and Costa Rica have shown by devoting some of their carbon tax revenues to forestry projects.[100]

Looking to Other Greenhouse Gases

To pre-empt the threat of seriously disruptive climate change, measures are also urgently needed to limit other climate-altering gases, which contribute less than carbon dioxide does to the warming now under way—approximately 36 percent of the total—but present special problems because of their powerful heat-trapping potential and/or long lifetimes in the atmosphere. Current concentrations of many of these gases are low but growing quickly, and have yet to receive adequate attention from policymakers. There are, however, many cost-effective ways in which to lower or, in some cases, completely eliminate these emissions.[101]

The second most important greenhouse gas, methane (CH_4), is 20 times as potent as carbon dioxide but is being emitted in far smaller quantities and accounts for 20 percent of the overall warming effect. Because methane lasts a short time in the atmosphere (approximately 12 years, compared with up to several hundred years for CO_2), its concentrations can be lowered relatively quickly. Since pre-industrial times, atmospheric CH_4 levels have increased by 145 percent, primarily from higher rates of fossil fuel extraction, agriculture, and waste disposal. Technologies are available that can capture as much as 90 percent of the methane emitted from coal mining and use it for energy, and are being demonstrated in China and several other countries. And the IPCC estimates that improved management of rice fields and livestock waste could cut global agricultural methane emissions by 35 percent. Landfill gas, meanwhile, can be used to generate electricity. The United States requires that large landfills capture and burn methane and provides a small tax credit for generating electricity this way; it projects a 60 percent cut in these emissions by 2000.[102]

Nitrous oxide (N_2O), another greenhouse gas, lasts 120 years in the atmosphere and is 200 times as potent as carbon dioxide. N_2O concentrations have increased 15 percent since pre-industrial times due to emissions from a variety of agricultural and industrial processes, including industrial fertilizer use and nylon production. Partially switching from manmade fertilizers to natural sources of nitrogen, such as human and plant waste, can cut N_2O emissions 17 percent from current rates while saving farmers money. Redirecting farm subsidies to encourage reduced fertilizer use—steps the United States and the European Union have tentatively begun to take—can accomplish this shift. Further, new processes are being developed for cutting N_2O emissions from nylon production—some by as much as 98 percent—and the United Kingdom projects it will lower these emissions by 95 percent between 1990 and 2000.[103]

Hydrofluorocarbons (HFCs) and perfluorocarbons (PFCs) are among the most potent greenhouse gases known. Introduced this decade, they are still produced in tiny quantities but their use is projected to grow significantly in coming years. HFCs, used as alternatives to ozone-depleting substances in refrigerants and cleaning and foaming agents, have between 1,300 and 24,000 times the warming effect of CO_2 and are projected to quadruple in use by 2050. PFCs, emitted from aluminum smelting and used in semiconductor manufacturing, have a warming effect between 6,500 and 9,200 times that of CO_2 and generally remain in the atmosphere for several thousand years.[104]

Most industrial countries have only made preliminary estimates of their HFC and PFC emissions, as few require that these be reported by industries. Many, however, project significantly greater future use of these substances. The United States restricts HFC and PFC use in applications where less-climate-changing substitutes can be used, and has established voluntary partnerships with aluminum and semiconductor companies to reduce emissions of PFCs, and develop substitutes for them. Together these measures are projected to avoid the equivalent of 23 million tons of carbon by 2010, twice the original estimate. Nevertheless, HFC and PFC emissions are projected to increase by 430 percent by 2020.[105]

Some European countries are beginning to address HFCs and PFCs. In fact, international controls of these substances have been advocated by Germany and the Netherlands since 1994. The Netherlands requires that industries report emissions of both substances and has set maximum leakage rates of HFCs from refrigerators. Denmark is experiencing only minor HFC use, and in 1996 announced it would phase out the chemical over the next decade. The

United Kingdom has signed voluntary agreements with 5 industry sectors to reduce HFC use. PFCs are already regulated in the U.K. aluminum industry—achieving a 65 percent cut from 1990 levels—and overall emissions in 2000 are expected to be below 1990 levels. European industries are also beginning to explore alternatives to these substances: in 1997 HFC producers proposed a voluntary agreement to ban the chemical's use in self-chilling drink cans.[106]

The most recent activity on HFCs and PFCs has taken place in Japan, whose climate plan was criticized by the climate treaty secretariat in 1996 for failing to include preliminary estimates of these emissions. In response, the government launched a two-year project to explore HFC and PFC alternatives and in July 1997 announced plans to regulate HFC use, calling on other nations to join its effort at Kyoto. As proposed, the regulations would limit HFC use to essential applications, in a closed loop, and with a recycling system for used HFCs. The following month, the government announced it would begin developing substitutes for PFCs, and would aim to have candidate chemicals ready within three years. Japanese semiconductor firms have in the meantime organized a task force with U.S. firms to coordinate efforts at phasing out PFC use.[107]

As efforts to tackle other greenhouse gases indicate, stabilizing the climate is a long-term endeavor covering a widening range of greenhouse gases—some of which may not be identified yet. It is essential to ensure that the replacements for these substances do not unwittingly contribute to other environmental problems—as occurred with the ozone layer protection effort. However, experience with the ongoing phaseout of ozone-depleting chemicals has also shown that ready alternatives to such chemicals can often be developed in a matter of years—not decades, as many industries have argued.[108]

The Challenge of Kyoto

A successful Kyoto Protocol could be an important step in the long-term effort to protect the world from the most devastating effects of climate change. But even at best, it will provide only the broadest framework—the targets, timetables, reporting requirements, and trading mechanisms needed to deal with climate change—leaving it to national governments to work out the details. The success of this endeavor will therefore hinge largely on the policies undertaken by individual countries to reduce their emissions of greenhouse gases. International coordination could make these changes more attractive politically, overcoming the competitiveness arguments that have stalled unilateral efforts and building trust between industrial and developing nations.

Many valuable policy lessons have been learned over the past few years that may point the way to solid progress in the post-Kyoto era. The climate itself may serve as a guide for responding to the challenge stabilization poses to humanity: to embark on an era of policy experimentation as profound and far-reaching as the geophysical experiment now under way. Just as the climate system is determined by the collective interaction of major components—the atmosphere, oceans, and biosphere—only a portfolio of mutually reinforcing policy measures is equal to the task of stabilizing it.[109]

Among the policies adopted so far, it is clear that the removal of energy subsidies has had the greatest short-term impact on emissions trends, in some cases contributing to sharp reductions. But this is largely a one-time effort, and many countries no longer have sizable energy subsidies to eliminate. Still, in countries such as Australia, Canada, China, Germany, and Russia, further cuts could have a great effect. And today's high subsidies for road use offer additional unrealized potential for lowering emissions. Experience shows that energy and emissions taxes can have a significant impact as well, but so far—with the partial exception of gasoline taxes—few countries have found the political courage to add new energy taxes that really bite.

Another useful lesson that can be drawn from the policy record to date is the proven effectiveness of energy efficiency standards, although the slow turnover of devices such as automobiles and home appliances means that it will take time for their full impact to be felt. Unfortunately, progress in setting new energy efficiency standards over the past five years has been limited. Governments have so far been most aggressive in pushing standards for buildings and appliances, but have been reluctant to adopt auto fuel economy standards in the face of strident industry opposition.

The Netherlands, meanwhile, has demonstrated that industry covenants can be effective, but only if they involve specific and ambitious goals, require reporting, are independently monitored, and include penalties for non-compliance—features that are mostly lacking in other countries' efforts to create similar programs. To make a real contribution to combating climate change, these agreements must also, as in the Netherlands, be treated as a complement to a broader framework of policies.

Some of the most innovative policies for climate change mitigation involve new incentive mechanisms used to encourage reliance on renewable energy and cogeneration. Countries such as Denmark, Germany, and India have shown that the right combination of tax incentives and generous purchase prices can spur private industry to invest large sums in these technologies. Although they are not yet in use on a sufficiently large scale to effect major reductions in carbon emissions, the rapid growth now under way in some renewable energy markets will allow them to make sizable contributions in coming decades.

As far as forests are concerned, it is now apparent that new sources of financing will need to be found if they are to help us balance the carbon budget. A growing number of developing countries have started to attract private investment for carbon sequestration, while a few industrial countries are beginning to draw upon public support for increasing forest area. One promising innovation that has emerged but is not yet widely employed is the partial use of carbon tax revenue to support forestry projects. Overall, carbon storage remains one of the most cost-effective yet least exploited means of slowing climate change.

Recent experience also suggests that rapid progress can be made in addressing other greenhouse gases. Incentives in a growing number of industrial and developing countries are beginning to encourage new technologies and processes to cut emissions of methane and nitrous oxide. HFCs and PFCs have yet to receive sufficient attention, but already there are signs that government regulations and required phaseouts could spur a speedy development of alternatives to these potent, long-lasting chemicals—much as the Montreal Protocol has sparked phasing out the use of ozone-depleting chemicals.

No single policy will by itself solve the climate problem.

One of the most important overall lessons to date of efforts taken to slow climate change is that no single policy will by itself solve the problem. Adjusting energy prices to accurately reflect the environmental consequences of fossil-fuel use is arguably the most important policy, but that reform alone will not suffice. Many market barriers—from lack of information to the anti-competitive practices of some industries—impede the implementation of well-established climate-friendly technologies and practices. Accordingly, only a diverse portfolio of policies can ultimately reverse the growth of emissions. Government policies that traditionally play out in isolated chambers must instead be assembled into complex orchestras, coordinating a wide range of instruments, and working in concert toward a shared goal.[110]

For 10 of the industrial countries to which the voluntary goal of stabilizing greenhouse emissions at the 1990 level applies, a clear hierarchy of the relative strengths and weaknesses of their climate policies emerges—making it possible to rank their comparative performance. (See Table 8.) Most successful are Denmark, Germany, and the Netherlands, each of which has developed a strong policy portfolio, though all have some gaps remaining. In this group, only Germany has actually cut its emissions, but this is in part due to the sharp cut in coal use in eastern Germany that resulted from reunification. All three countries appear likely to reduce their emissions during the next decade. The European Union as a whole is also playing a useful role in urging its 15 member states to consider a range of new policies, including a tax on energy and carbon, and to coordinate their actions.[111]

France, Japan, Sweden, and the United Kingdom form a middle group of countries that are making some progress but still have major holes in their policy mix. Japan, for example, now has a strong solar incentive program, but has weak efficiency standards. The United Kingdom has made good progress in reducing subsidies, but has a weak renewables policy. With further advances in their policies, these four nations have a good chance of reducing their relatively modest per capita emissions in the near future.

On the next rung of the climate policy ladder is the United States. The country's low energy taxes have not yet been raised, and auto efficiency standards have in effect been permitted to lapse. U.S. renewables incentives are too weak to overcome the negative effects of electric utility restructuring, and voluntary industry programs have been weak and ineffectual. Ironically, U.S. policy was doing more to limit carbon emissions a decade ago, before the greenhouse problem was even officially recognized. However, the United States still stands out for its strong appliance efficiency standards, and a number of state and local governments are experimenting with promising policies in support of renewable energy and public transit.[112]

The weakest climate policies to date are found in Australia and Canada, countries built on extractive industries with low fossil fuel prices. Neither has done much to reduce its substantial producer subsidies for fossil fuels, and

TABLE 8

Climate Policies, Selected Industrial Countries, 1997

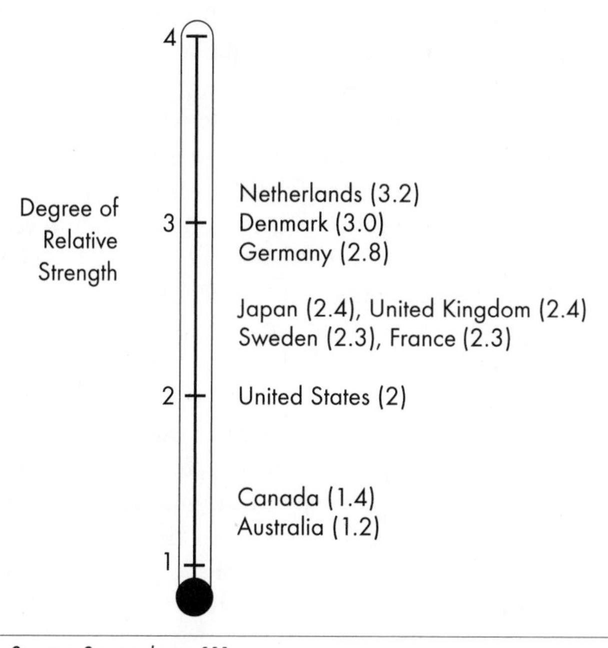

Source: See endnote 111.

both continue to have low fuel taxes. Despite having huge renewable resources, both countries have allowed their monopoly power companies to almost completely block renewables from the power grid. Some of their states and provinces have taken creative steps to promote energy-efficient cars and renewable energy, but these have so far worked at cross-purposes with national policies. Without major changes, emissions from Australia and Canada are likely to continue upward for the indefinite future.[113]

Given the criticism that has been leveled at the climate convention by some politicians for not including developing countries in the year 2000 goals agreed to in Rio, it is useful to consider the actual record of large developing nations such as Brazil, China, and India. By comparison with the industrial countries assessed here, these nations are not doing so badly. Each of the three has implemented meaningful policy reforms in the past decade, including politically difficult reductions in fossil fuel subsidies, with noticeable results in increased renewables use in Brazil and India, and improved efficiency in China.

This brief history of climate policy, with its short but instructive record of successes and failures, makes it clear that it is far too early to despair for the future of the climate treaty. So, too, does experience with past international environmental agreements, which shows that governments are often slow to implement new treaties at first, but can then move into high gear and make rapid progress. The Vienna Convention to protect the ozone layer was originally passed in 1985, but it took two years to achieve a legally binding protocol, and several more years before steep declines in chlorofluorocarbon (CFC) production occurred as the protocol was strengthened by subsequent amendments. The climate convention is on a slower track, but that is hardly surprising, given the central role that fossil fuels play in today's economy, and the power of the industries that supply and use them.[114]

Where do we go from here? The Third Conference of the Parties to the climate convention in Kyoto presents a unique opportunity to strengthen the convention and spur governments to bolster their climate policies. The European Union (under Dutch leadership) has proposed that the Kyoto Protocol include a mandatory requirement that industrial countries lower their emissions to 7.5 percent below 1990 levels by 2005, and 15 percent below 1990 levels by 2010—targets that would push most industrial countries to substantially strengthen their climate policies. Japan has proposed a less ambitious 5 percent emissions reduction by 2010. Weaker still is the United States proposal for a return to 1990 emissions levels by 2008–2012—a 10-year extension of the goal agreed to in Rio. The striking correlation between these stances and the tiers of our ranking suggests that the governmental negotiating positions in Kyoto are less a reflection of their technical and economic potential for reducing greenhouse gas emissions than an indication of their current domestic commitment to addressing climate change.[115]

The long-term significance of the targets and timetables agreed to in Kyoto may lie in their ability to unleash the ingenuity needed to address climate change. The tougher the targets, the greater the motivation for policy change. But the ultimate effectiveness of the climate convention may depend on the bargain that is struck between the industrial countries that contribute most heavily to today's emissions and the developing nations that will increasingly dominate emissions in the next century. The challenge is to phase the latter gradually into the treaty's emissions limits—without threatening their development prospects. Unfortunately, the diplomatic well has been soured by the failure of industrial countries to provide the financial assistance that was promised at Rio—particularly through the Global Environment Facility (GEF), the funding mechanism established to help developing countries address climate change and other global environmental problems.[116]

Cutting a deal will thus require solving several equations simultaneously. One option is for developing countries such as South Korea and Saudi Arabia, which now have per capita emissions as high as those of many industrial countries, to commit to slowing their emissions growth. In addition, it is essential that industrial nations come through with meaningful funding for technology transfer so as to persuade poorer developing nations of the economic advantages of climate protection. If industrial countries gained confidence that developing countries would join them in the move away from fossil fuels, they might be more ready to embrace stronger emissions limits. But developing countries can only proceed with that transition if industrial countries are actively commercializing the needed technologies.

The kernels of a successful bargain can be found in the 1987 Montreal Protocol to protect the ozone layer, whose Interim Multilateral Fund served as a model for the GEF. In the London Amendment to the Protocol, the Fund was established to provide financial assistance to developing countries for the transfer of non-ozone-depleting technologies. With this assistance, developing countries agreed to limit their emissions, but on a slower timetable than industrial countries; despite this delay, many have now phased out their CFC production ahead of schedule. In addition to developing-country participation, the Montreal Protocol also achieved the cooperation of industry groups and the United Nations in providing information on alternative technologies to developing countries.[117]

Such institutional innovations could help the world resolve the unprecedented diplomatic complexity posed by the climate problem. They might include an equitable phase-in of stronger developing-country commitments through an accelerated transfer of funds and technology; widespread industry involvement in disseminating information on carbon-free technologies to the developing world's nascent industries; and a heightened role for the U.N. and the World Bank in spearheading technological cooperation on climate abatement technologies. The World Bank presently funds a large number of fossil fuel projects that contribute to climate change, but Bank officials have recently expressed an interest in funding innovative projects that reduce emissions. Prime Minister Ryutaro Hashimoto of Japan proposed at the June 1997 Special Session of the U.N. General Assembly that industrial countries collectively commit to a Green Initiative for channeling aid and technologies to the developing world. A wide range of industries, from insurance to sustainable energy, have become active in the climate negotiations and share an enormous interest in helping deploy clean technologies in the developing world. And UNEP's Environmental Technology Centre, based in Tokyo, is well positioned to take on the role of clearinghouse for North-South technological cooperation related to climate change.[118]

Part of the needed transfers of finances and technology could also take place through an emissions budget and trading system now being considered under the protocol. Under such a program, countries would be allowed to purchase emissions permits from other countries to stay within their limits—encouraging them to seek out reductions where it is cheapest to do so. Such a system has proven effective at the domestic level in the United States in reducing emissions of lead, CFCs, and sulfur dioxide well ahead of schedule and at a much lower cost than originally anticipated. The U.S. sulfur dioxide emissions permit trading program, for example, has helped reduce emissions by 30 percent since 1994 at less than one tenth the cost projected by industry.[119]

The creation of a viable international trading regime for greenhouse gases faces far more complicated political and institutional challenges, and will take many years to develop fully. Moreover, experience from past trading systems shows that binding emissions limits must be stringent to make such a system effective, and it remains to be seen if the Kyoto target will meet this criterion. If the limits in Kyoto are strong enough, however, such a regime may over time stimulate the transfer of climate-friendly technologies and encourage countries to act early in reducing domestic emissions. In fact, if properly designed and managed, a tradeable permit scheme could serve to focus the attention of *all* nations on opportunities for cost-effectively limiting their own emissions.[120]

Developing countries can only proceed if industrial countries commercialize the needed technologies.

The ultimate political test, then, will take place when government officials return home from Kyoto. Recent experience provides useful guidelines as national governments take steps to implement the Kyoto Protocol. But action at all levels—by state and local governments, private businesses, and concerned citizens—will be needed to make the protocol work.[121]

Human innovation created the climate problem, and a new burst of innovation—both technological and political—will be needed to solve it. The challenge is to find a new and cleaner way to increase economic prosperity and to do so soon—before it is too late. If delegates can keep this goal in mind as they leave the International Conference Hall in Kyoto, the city that was spared from the devastation of the atomic bomb during World War II may one day also be remembered as the city that helped spare the world from dangerous climate change.

Notes

1. Worldwatch estimate based on T.A. Boden, G. Marland, and R.J. Andres, *Estimates of Global, Regional and National Annual CO_2 Emissions From Fossil Fuel Burning, Hydraulic Cement Production, and Gas Flaring: 1950-92* (Oak Ridge, TN: Carbon Dioxide Information Analysis Center, Oak Ridge National Laboratory, December 1995) and British Petroleum, *BP Statistical Review of World Energy 1997* (London: Group Media & Publications, 1997); Stephen H. Schneider, *Laboratory Earth: The Planetary Gamble We Can't Afford to Lose* (New York: BasicBooks, 1997); Timothy Whorf and C.D. Keeling, Scripps Institution of Oceanography, La Jolla, CA, letter to authors, 10 February 1997; J.T. Houghton et al., eds., *Climate Change 1995: The Science of Climate Change*, Contribution of Working Group I to the Second Assessment Report of the Intergovernmental Panel on Climate Change (IPCC) (Cambridge, U.K.: Cambridge University Press, 1996).

2. Ibid; Robert T. Watson, Marufu C. Zinyowera, and Richard H. Moss, eds., *Climate Change 1995: Impacts, Adaptation, and Mitigation*, Contribution of Working Group II to the Second Assessment Report of the Intergovernmental Panel on Climate Change (Cambridge, U.K.: Cambridge University Press, 1996); Thomas R. Karl et al., "The Coming Climate," *Scientific American*, May 1997; Houghton et al., op. cit. note 1.

3. Watson et al., op. cit. note 2.

4. *United Nations Framework Convention on Climate Change, Text* (Geneva: U.N. Environment Programme (UNEP)/World Meteorological Organization (WMO), Information Unit on Climate Change, 1992); Worldwatch estimate based on Boden et al., op. cit. note 1, and BP, op. cit. note 1.

5. Christopher Flavin and Nicholas Lenssen, *Power Surge: Guide to the Coming Energy Revolution* (New York: W.W. Norton & Company, 1994); Watson et al., op. cit. note 2.

6. British Information Services, "UK Will Take the Lead in Fight Against Global Warming," 4 June 1997.

7. James P. Bruce, Hoesung Lee, and Erik F. Haites, eds., *Climate Change 1995: Economic and Social Dimensions of Climate Change* (Cambridge, U.K.: Cambridge University Press, 1996).

8. Watson et al., op. cit. note 2; A.J. McMichael et al., eds., *Climate Change and Human Health*, An assessment prepared by a task group on behalf of the World Health Organization (WHO), WMO, and UNEP (Geneva: WHO, 1996).

9. Ross Gelbspan, *The Heat Is On: The High Stakes Battle over Earth's Threatened Climate* (New York: Addison-Wesley Publishing Company, Inc.,

1997); Fred Pearce, "Greenhouse Wars," *New Scientist,* 19 July 1997; Houghton et al., op. cit. note 1; Jerry D. Malhman, "Anticipated Climate Changes in a High-CO_2 World: Implications for Long-Term Planning," U.S. Global Change Research Program Seminar, Washington, D.C., 15 September 1997; Mike McCartney, "Is the ocean at the helm?" *Nature,* 7 August 1997; Stephan Rahmstorf, "Risk of sea change in the Atlantic," *Nature,* 28 August 1997.

10. M.L. Parry, T.R. Carter, and M. Hulme, "What is dangerous climate change?" *Global Environmental Change,* no. 1, 1996; Christian Azar and Henning Rodhe, "Targets for Stabilization of Atmospheric CO_2," *Science,* 20 June 1997.

11. National Institute of Public Health and the Environment (RIVM), *Linking Near-Term Action to Long-Term Climate Protection: The Safe Landing Analysis* (The Netherlands, February 1996); Watson et al., op. cit. note 2; carbon emissions data, including Figure 1, from Worldwatch estimate based on BP, op. cit. note 1; Bruce et al., op. cit. note 7; Ozone Action, *The Case for Early Action: An Ozone Action Report on Global Climate Change* (Washington, D.C., September 1997); Stephen H. Schneider and Lawrence H. Goulder, "Achieving low-cost emissions targets," *Nature,* 4 September 1997.

12. Figure 2 from Worldwatch estimate based on Boden et al., op. cit. note 1, and BP, op. cit. note 1; *U.N. Framework Convention,* op. cit. note 4; Worldwatch estimate based on Boden et al., op. cit. note 1, and BP, op. cit. note 1.

13. Watson et al., op. cit. note 2; Worldwatch estimate based on American Automobile Manufacturers Association, *World Motor Vehicle Data, 1997 Edition* (Detroit, 1997); John Griffiths, "Car Numbers to Double and Threaten the Environment," *Financial Times,* 15 February 1996; Lee Schipper, "The Carnival Is Over," *Tomorrow,* September/October 1997; International Energy Agency (IEA), *Indicators of Energy Use and Efficiency: Understanding the link between energy and human activity* (Paris: Organisation for Economic Co-operation and Development (OECD)/IEA, 1997).

14. Worldwatch estimate based on Boden et al., op. cit. note 1, and BP, op. cit. note 1.

15. Ibid; Worldwatch estimate based on Boden et al., op. cit. note 1, and BP, op. cit. note 1.

16. Figure 3 from ibid; Figure 4, Tables 1 and 2, and per capita estimate from Worldwatch estimate based on Boden et al., op. cit. note 1, BP, op. cit. note 1, Population Reference Bureau (PRB), *World Population Data Sheet* (Washington, D.C., 1997), and World Bank, *The World Bank Atlas 1997* (Washington, D.C., 1997); IEA, *World Energy Outlook: 1996 Edition* (Paris: OECD/IEA, 1996).

17. United States Senate, *Senate Resolution 98* (Washington, D.C., 25 July 1997); William K. Stevens, "U.S. and Japan Key to Outcome in Climate Talks," *New York Times*, 12 August 1997.

18. Holdren quote from White House Conference on Climate Change, Georgetown University, Washington, D.C., 6 October 1997; Vicki Norberg-Bohm and David Hart, "Technological Cooperation: Lessons from Development Experience," in Henry Lee, ed., *Shaping National Responses to Climate Change* (Washington, D.C.: Island Press, 1996); Watson et al., op. cit. note 2.

19. Flavin and Lenssen, op. cit. note 5; Ernst von Weizsacker, Amory B. Lovins, and L. Hunter Lovins, *Factor Four: Doubling Wealth, Halving Resource Use, The New Report to the Club of Rome* (London: Earthscan Publications Limited, 1997).

20. Michiyo Nakamoto and John Griffiths, "Toyota car will run on petrol or batteries," *Financial Times*, 26 March 1997; "Japan's automakers push low-emission vehicles," *Daily Yomiuri*, 8 April 1997; Rocky Mountain Institute, "Hypercars: The Next Industrial Revolution," paper presented at the 13th International Electric Vehicle Symposium, Osaka, Japan, October 1996.

21. Gerald L. Cler and Michael Shepard, "Distributed Generation: Good Things Are Coming in Small Packages," *E Source Tech Update*, November 1996; Sallie L. Gaines, "Pint-Size Producer of Power Unveiled," *Chicago Tribune*, 4 June 1997.

22. Watson et al., op. cit. note 2; von Weizsacker et al., op. cit. note 19.

23. Wind power data, including Figure 5, from Christopher Flavin, "Wind Power Growth Continues," in Lester R. Brown, Michael Renner, and Christopher Flavin, *Vital Signs 1997* (New York: W.W. Norton & Company, 1997); Flavin and Lenssen, op. cit. note 5; 1997 figure is preliminary Worldwatch estimate.

24. Solar power data, including Figure 6, from Molly O'Meara, "Solar Cell Shipments Keep Rising," in Brown et al., op. cit., note 23; Flavin and Lenssen, op. cit. note 5; 1997 figure is preliminary Worldwatch estimate.

25. Jesse H. Ausubel, "Technical progress and climatic change," *Energy Policy*, April/May 1995; Nebojsa Nackicenovic, "Freeing Energy From Carbon," in Jesse H. Ausubel and H. Dale Langford, eds., *Technological Trajectories and the Human Environment* (Washington, D.C.: National Academy Press, 1997).

26. Jasinowski quote from *Morning Edition*, National Public Radio, 10 September 1997; Robert Repetto and Duncan Austin, *The Costs of Climate Protection: A Guide for the Perplexed* (Washington, D.C.: World Resources

Institute (WRI), 1997); W. Brian Arthur, *Increasing Returns and Path Dependence in the Economy* (Ann Arbor, MI: The University of Michigan Press, 1994); Michael Grubb, Thierry Chapuis, and Minh Ha Duong, "The economics of changing course: Implications of adaptability and inertia for optimal climate policy," *Energy Policy*, April/May 1995; Michael Grubb, "Technologies, energy systems and the timing of CO_2 emissions abatement," *Energy Policy*, February 1997; John S. Hoffman, U.S. Environmental Protection Agency (EPA), "The potential of institutional, organizational, and technological change to improve the future productivity of the energy economy," discussion paper (Washington, D.C., 14 June 1996).

27. Florentin Krause, "The costs of mitigating carbon emissions: a review of methods and findings from European studies," *Energy Policy*, October/November 1996; Florentin Krause, "The Costs and Benefits of Cutting U.S. Carbon Emissions: A Critical Review of the Economic Arguments of the Fossil Fuel Lobby" (El Cerrito, CA: International Project for Sustainable Energy Paths, May 1997).

28. U.S. Department of Energy (DOE), *Scenarios of U.S. Carbon Reductions: Potential Impact of Energy Technologies by 2010 and Beyond*, Prepared by the Interlaboratory Working Group on Energy-Efficient and Low-Carbon Technologies (Washington, D.C., 25 September 1997).

29. Shell International Limited, *The Evolution of the World's Energy Systems* (London, 1996).

30. World Wide Fund for Nature (WWF), *Policies and Measures to Reduce CO_2 Emissions by Efficiency and Renewables*, prepared by Department of Science, Technology and Society, Utrecht University (The Netherlands, 1996); Alliance to Save Energy et al., *Energy Innovations: A Prosperous Path to a Clean Environment* (Washington, D.C., June 1997); Sierra Club of Canada, *Rational Energy Program: Analysis of the Impact of Rational Measures to the Year 2010* (Ottawa, September 1996); "Sierra Club's Climate Change Proposals Sound, According to Government Analysis," *International Environment Reporter*, 2 October 1996; Environment Agency of Japan, *Summary of the Report of the Study Group on Global-Warming Abatement Technologies* (Tokyo, May 1996); Australia Institute, "Climate Change Policies in Australia," briefing to a meeting of the Ad Hoc Group on the Berlin Mandate, Bonn, 5 August 1997.

31. Watson et al., op. cit. note 2; Bruce et al., op. cit. note 7.

32. Laurie Michaelis, "Reforming Coal and Electricity Subsidies," Working Paper 2, Annex I Expert Group on the UN FCCC (Paris: OECD, July 1996); Robert T. Watson, Marufu C. Zinyowera, and Richard H. Moss, eds., *Technologies, Policies and Measures for Mitigating Climate Change*, IPCC Technical Paper 1 (Geneva, November 1996); Bruce et al., op. cit. note 7.

33. World Bank, *Expanding the Measure of Wealth* (Washington, D.C., June 1997); IEA, *Climate Change Policy Initiatives, 1995-96 Update, Volume II,*

Selected Non-IEA Countries (Paris: OECD/IEA, 1996); Climate Action Network-Central and Eastern Europe (CAN-CEE) and Climate Network Europe (CNE), *Independent NGO Evaluations of National Plans for Climate Change Mitigation, Second Review* (Brussels: CNE, November 1996); Worldwatch estimate based on Boden et al., op. cit. note 1, and BP, op. cit. note 1; World Bank, op. cit. this note; Worldwatch estimate based on Boden et al., op. cit. note 1, and BP, op. cit. note 1; Michaelis, op. cit. note 32; Andre de Moor and Peter Calamai, *Subsidizing Unsustainable Development: Undermining the Earth with public funds* (San José, Costa Rica: Earth Council and Institute for Research on Public Expenditure, 1997).

34. Greenpeace International, *Energy Subsidies in Europe*, prepared by Institute for Environmental Studies, Vrije University (Amsterdam, May 1997); Michaelis, op. cit. note 32, U.K. Secretary of State for the Environment, *Climate Change: The U.K. Programme, United Kingdom's Second Report under the Framework Convention on Climate Change* (London: The Stationery Office Limited, February 1997); U.N. Framework Convention on Climate Change, "Report on the in-depth review of the national communication of the United Kingdom" (Bonn, 24 February 1997); IEA, *Energy Policies of IEA Countries, 1996 Review* (Paris: OECD/IEA, 1996); Worldwatch estimate based on BP, op. cit. note 1; Worldwatch estimate based on Boden et al., op. cit. note 1, and BP, op. cit. note 1.

35. World Bank, op. cit. note 33; Michaelis, op. cit. note 32; David Malin Roodman, *Paying the Piper: Subsidies, Politics, and the Environment*, Worldwatch Paper 133 (Washington, D.C.: Worldwatch Institute, December 1996); Climate Network Europe (CNE) and US Climate Action Network (US CAN), *Independent NGO Evaluations of National Plans for Climate Change Mitigation, OECD Countries, Fifth Review* (Washington, D.C.: US CAN, October 1997); William G. Mahoney, "German Subventions at Record High; Coal Supports Blamed for Cost Hike," *The Solar Letter*, 12 September 1997.

36. De Moor and Calamai, op. cit. note 33; Michaelis, op. cit. note 32; IEA, op. cit. note 34; Friends of the Earth, *Green Scissors '97*, May 1997; de Moor and Calami, op cit. note 33; CNE and US CAN, op. cit. note 35.

37. Michaelis, op. cit. note 32; Tim O'Riordan and Jill Jager, eds., *Politics of Climate Change: A European Perspective* (London: Routledge, 1996); OECD, *Subsidies and Environment: Exploring the Linkages* (Paris, 1996).

38. IEA, op. cit. note 33; subsidy removal data, including Table 3, from World Bank, op. cit. note 33; China projection from World Bank, *China: Issues and Options in Greenhouse Gas Emissions Control*, Summary Report (Washington, D.C., 1994).

39. De Moor and Calamai, op. cit. note 33; Laurie Michaelis, "Sustainable Transport Policies: CO_2 Emissions From Road Vehicles," Working Paper 1, Annex I Expert Group on the UN FCCC (Paris: OECD, July 1996); Jane Holtz Kay, *Asphalt Nation: How the Automobile Took Over America and How*

We Can Take It Back (New York: Crown Publishers, 1997); Figure 7 from Stacy C. Davis et al., *Transportation Energy Data Book: Editions 12, 16, and 17* (Oak Ridge, TN: Oak Ridge National Laboratory, 1992, 1996, 1997); de Moor and Calamai, op. cit. note 33.

40. Laurie Michaelis, "Policies and Measures to Encourage Innovation in Transport Behavior and Technology," Working Paper 13, Annex I Expert Group on the UN FCCC (Paris: OECD, March 1997); European Conference of Ministers of Transport (ECMT), "Report on the Monitoring of Policies for Reduction of CO_2 Emissions" (Paris, 24 March 1997); OECD, "Criteria for Sustainable Transport" (Paris, July 1996); John Pucher and Christian Lefevre, *The Urban Transport Crisis in Europe and North America* (London: MacMillan Press Ltd., 1996); U.N. Framework Convention, "Report on the in-depth review of the national communication of Denmark" (Bonn, 6 December 1996); Dean Anderson, Michael Grubb, and Joanna Depledge, *Climate Change and the Energy Sector: A country-by-country analysis of national programmes, Volume 1: The European Union* (London: Financial Times Energy, 1997); Ministry of Housing, Spatial Planning, and the Environment, *Second Netherlands' National Communication of Climate Change Policies*, Prepared for the Conference of the Parties under the Framework Convention on Climate Change (The Hague, April 1997); U.N. Framework Convention, "Report on the in-depth review of the national communication of the Netherlands" (Bonn, 31 July 1996); CNE and US CAN, op. cit. note 35.

41. Michaelis, op. cit. note 40; ECMT, op. cit. note 40; ECMT, *Sustainable Transport in Central and Eastern European Cities* (Paris: OECD, 1996); Michaelis, op. cit. note 40.

42. Richard Baron, "Economic Fiscal Instruments: Taxation," Working Paper 3, Annex I Expert Group on the UN FCCC (Paris: OECD, July 1996); Watson et al., op. cit. note 32.

43. Frank Muller, "Mitigating Climate Change: The Case for Energy Taxes," *Environment*, March 1996; OECD, *Environmental Taxes in OECD Countries* (Paris, 1996); Table 4 from Baron, op. cit. note 42; Ministry of Housing, op. cit. note 40; Baron, op. cit. note 42; U.N. Framework Convention, "Report on...Denmark," op. cit. note 40.

44. Muller, op. cit. note 43; Baron, op. cit. note 42; "EC tax reform group revives carbon/energy tax," *ENDS Report 261*, October 1996; "Carbon Taxes, More Energy Efficiency Urged by Environment Agency to Avert Global Crisis," *International Environment Reporter*, 11 June 1997.

45. IEA, op. cit. note 33; CAN-CEE and CNE, op. cit. note 33; Thomas E. Lovejoy, "Lesson From a Small Country," *Washington Post*, 22 April 1997.

46. Gasoline tax and price data, including Figure 7, from Michaelis, op. cit. note 40; ECMT, op. cit. note 40; U.K. Secretary of State for the Environment, op. cit. note 34; *Economist*, "Hot Air?" 9 August 1997.

47. Watson et al., op. cit. note 32; ECMT, op. cit. note 40; U.N. Framework Convention, "Report on the in-depth review of the national communication of Austria" (Bonn, 10 December 1996).

48. Watson et al., op. cit. note 32; IEA, op. cit. note 13; IEA, *Energy Efficiency and Climate Change Response* (Paris, June 1996); Watson et al., op. cit. note 32.

49. Figure 8 from ibid; Michaelis, op. cit. note 39; Watson et al., op cit. note 32.

50. Automobile fuel efficiency data, including Table 5, from Michaelis, op. cit. note 39; Public Citizen, Critical Mass Energy Project, press release, 22 May 1997; Matthew L. Wald, "U.S. Increasing Its Dependence on Oil Imports," *New York Times*, 11 August 1997.

51. IEA, *Voluntary Actions for Energy-Related CO_2 Abatement* (Paris: OECD/IEA, 1997); Michaelis, op. cit. note 39; "European Parliament Seeks Fuel Economy Standards," *Global Environmental Change Report*, 25 April 1997; John Duffy, *Energy Labeling, Standards and Building Codes: A Global Survey and Assessment for Selected Developing Countries* (Washington, D.C.: International Institute for Energy Conservation (IIEC), March 1996).

52. Watson et al., op. cit. note 32.

53. IEA, op. cit. note 51; CNE and US CAN, op. cit. note 35; von Weizsacker et al., op. cit. note 19; U.N. Framework Convention, "Report on...the Netherlands," op. cit. note 40; idem, "Report on the in-depth review of the national communication of Germany" (Bonn, 21 July 1997); idem, "Report on the in-depth review of the national communication of Japan" (Bonn, 28 June 1996).

54. U.N. Framework Convention, "Report on the in-depth review of the national communication of Canada" (Bonn, 21 February 1996); National Air Issues Coordinating Committee (NAICC), *1996 Review of Canada's National Action Program on Climate Change* (Ottawa, November 1996); Duffy, op. cit. note 51; U.S. Department of State, *Second U.S. National Communication, Submitted Under the United Nations Framework Convention on Climate Change*, draft (Washington, D.C., 9 May 1997); Eric Jay Dolin, "EPA's Voluntary Pollution Prevention at a Profit," *Ecological Economics Bulletin*, spring 1997; U.S. Department of State, op. cit. this note.

55. Duffy, op. cit. note 51; "Tax Incentives for Energy-Saving Goods, More Building of Efficient Housing Planned," *International Environment Reporter*, 2 October 1996.

56. Watson, op. cit. note 32; Fiona Mullins, "Demand Side Efficiency: Energy Efficiency Standards for Traded Products," Working Paper 5, Annex I Expert Group on the UN FCCC (Paris: OECD, July 1996); IEA, op. cit. note 48; Mullins, op. cit. this note; Watson, op. cit. note 32; U.S. Department of State, op. cit. note 54; CNE and US CAN, op. cit. note 35; Mullins, op. cit. this note.

57. Ibid; Duffy, op. cit. note 51; IIEC, *Examples in Action: Sustainable Energy Experiences in Developing and Transition Countries* (Washington, D.C., 1996).

58. Mullins, op. cit. note 56; "More Power Consumption, Computer Use, Larger Cars Implicated in CO_2 Increase," *International Environment Reporter*, 10 July 1996; Global Environment Information Center, "Choice by CO_2" (Tokyo, 1997).

59. Watson et al., op. cit. note 32; William R. Moomaw, "Industrial emissions of greenhouse gases," *Energy Policy*, October/November 1996.

60. IEA, op. cit. note 51; Mark Storey, "Demand Side Efficiency: Voluntary Agreements with Industry," Working Paper 8, Annex I, Expert Group on the UN FCCC (Paris: OECD, December 1996).

61. Ministry of Housing, op. cit. note 42; Table 6 from IEA, op. cit. note 51; Storey, op. cit. note 60; U.N. Framework Convention, "Report on ...the Netherlands," op. cit. note 40; "Pacts Would Aim to Make Dutch Industries Most Energy-Efficient Performers in World," *International Environment Reporter*, 25 June 1997.

62. Bundesverband der Deutschen Industrie, *Updated and Extended Declaration by German Industry and Trade on Global Warming Prevention* (Cologne, July 1996); idem, *CO_2 Monitoring* (Cologne, 26 February 1996); Storey, op. cit. note 60; IEA, op. cit. note 51; Klaus Rennings et al., *Voluntary Agreements in Environmental Protection—Experiences in Germany and Future Perspectives*, Discussion Paper No. 97-04 E (Mannheim, Germany: Zentrum für Europaische Wirtschaftforschung, March 1997); European Environment Agency, *Environmental Agreements: Environmental Effectiveness* (Copenhagen, 1997).

63. U.S. Department of State, op. cit. note 54; IEA, op. cit. note 51; Storey, op. cit. note 60; U.N. Framework Convention, "Report on the in-depth review of the national communication of the United States of America" (Bonn, 26 February 1996); U.S. Department of State, op. cit. note 54.

64. Ibid; Dolin, op. cit. note 54; U.S. General Accounting Office, *Global Warming: Information on the Results of Four of EPA's Voluntary Climate Change Programs*, Report to Congressional Committees (Washington, D.C., June 1997); Peter Tulej, "A Bright Light in Poland," *E-Notes*, August 1997; Ming Yang and Peter du Pont, "The Green Lights of China," *E-Notes*, March 1997; Steve Nadel, Guan Fu Min, Yu Cong, and Hu Dexia, *Lighting Efficiency in*

China: Current Status and Future Directions (Washington, D.C.: American Council for an Energy-Efficient Economy, March 1997).

65. IEA, op. cit. note 51; Storey, op. cit. note 60; NAICC, op. cit. note 54; Pembina Institute for Appropriate Development, *Corporate Action on Climate Change 1996: An Independent Review* (Drayton Valley, Alberta, April 1997); Australia Institute, op. cit. note 30.

66. IEA, op. cit. note 51; "Japanese Industry Vows CO_2 Reductions," *Global Environmental Change Report*, 14 March 1997; Storey, op. cit. note 60; "MITI Asks Industry to Aim for Improvements in Energy Efficiency of 1 Percent Annually," *International Environment Reporter*, 19 March 1997; U.N. Framework Convention, "Report on...Japan," op. cit. note 53; Jonathan Lloyd-Owen, "Climate Change Fence-Sitting," *Tomorrow*, October 1997.

67. Bob Price and Caroline Hazard, "Voluntary Programs Catalyze Efficiency," *E-Notes*, March 1997; IIEC, op. cit. note 57.

68. Watson, op. cit. note 32; IEA, op. cit. note 51.

69. Watson, op. cit. note 32; Bruce et al., op. cit. note 7.

70. Text and Figure 9 from IEA, *IEA Energy Technology R&D Statistics, 1974–1995* (Paris, 1997); Microsoft data from "Microsoft Corporation Information," as posted at http://www.microsoft.com/corpinfo/about-ms.htm, 24 October 1997.

71. Text and Table 7 from IEA, *Renewable Energy Policy in IEA Countries* (Paris, 1997).

72. Ibid.

73. Ibid.

74. U.S. Office of Technology Assessment (OTA), *Renewing Our Energy Future* (Washington, D.C.: Government Printing Office, September 1997); IEA, op. cit. note 71; Ministry of the Environment, Denmark, *Climate Protection in Denmark, National Report of the Danish Government in Accordance with Article 12 of the United Nations Convention on Climate Change* (Copenhagen, 1994); "Danish Turbine Industry Record," *Wind Directions*, October 1997; WWF, op. cit. note 30; COGEN Europe, *European Cogeneration Review 1997* (Brussels, May 1997); U.N. Framework Convention, "Report on...Denmark," op. cit. note 40.

75. OTA, op. cit. note 74; IEA, op. cit. note 71; Jan Hamrin and Nancy Rader, *Investing in the Future: A Regulator's Guide to Renewables* (Washington, D.C.: National Association of Regulatory Utility Commissioners, 1993); California Energy Commission, Sacramento, CA, letter to authors, 17 October 1997.

76. OTA, op. cit. note 74; IEA, op. cit. note 71; Flavin, op. cit. note 23.

77. COGEN Europe, op. cit. note 74; WWF, op. cit. note 30; U.N. Framework Convention, "Report on...the Netherlands," op. cit. note 40; Ministry of Housing, op. cit. note 40; "New Skyline for Dutch coast," *Environmental News from the Netherlands*, February 1997; Ministry of Housing, op. cit. note 40.

78. IEA, op. cit. note 71; OTA, op. cit. note 74; Natural Resources Canada, *Renewable Energy Strategy: Creating a New Momentum* (Ottawa, October 1996); IEA, op. cit. note 71; OTA, op. cit. note 74.

79. IEA, op. cit. note 71; U.N. Framework Convention, "Report on...Germany," op. cit. note 53; Andreas Wagner, *Feed-In Tariffs for Renewable Energies in Europe—An Overview* (Bonn: European Association for Solar Energy, September 1997); Flavin, op. cit. note 23; CNE and US CAN, op. cit. note 35.

80. Wagner, op. cit. note 79; IEA, op. cit. note 71; Flavin, op. cit. note 23.

81. IEA, op. cit. note 71; U.N. Framework Convention, "Report on...the United Kingdom," op. cit. note 34; U.K. Secretary of State for the Environment, op. cit. note 34.

82. OTA, op. cit. note 74; IEA, op. cit. note 71; The Ministry of Foreign Affairs, Government of Japan, "Global Environmental Problems—Japanese Approaches," distributed at the Special Session of the U.N. General Assembly, New York, June 1997; O'Meara, op. cit. note 24.

83. "Japan Plans Big Solar Buildup," *Finance in the Greenhouse*, November 1996; Dean Anderson, Michael Grubb, and Joanna Depledge, *Climate Change and the Energy Sector: A country-by-country analysis of national programmes, Volume 2: Non-EU OECD Countries* (London: Financial Times Energy, 1997); Brenda Biondo, "Pushing Solar Energy Through the Roof," *Solar Industry Journal*, second quarter 1997; "Japan to boost subsidies for clean energy use," *Gallon Environment Letter*, 3 September 1997.

84. OTA, op. cit. note 74; Biondo, op. cit. note 83; "German PV Program Applications-Centered," *PV News*, September 1997.

85. IEA, op. cit. note 71; OTA, op. cit. note 74; White House, "President Clinton's Address to UN General Assembly Special Session," New York, 26 June 1997; Yih-huei Wan, *Net Metering Programs*, Topical Issues Brief, Prepared for National Renewable Energy Laboratory (NREL) (Boulder, CO, December 1996); idem, letter to authors, 9 September 1997.

86. IEA, op. cit. note 71; Seth Dunn, "Power of Choice," *World Watch*, September/October 1997.

87. IIEC, op. cit. note 57; Government of Brazil, "The Brazilian Fuel Ethanol Program," distributed at the Special Session of the U.N. General Assembly, New York, June 1997; Mary Milliken and Mara Lemos, "Brazil is Trying to Save Invention, Alcohol Power," *New York Times*, 27 June 1997.

88. IIEC, op. cit. note 57; Flavin, op. cit. note 23; Rakesh Bakshi, "Country Survey: India," *Wind Directions*, April 1997.

89. World Bank, *Renewable Energy for Electric Power* (Washington, D.C.,1996); Flavin, op. cit. note 23; O'Meara, op. cit. note 24; IIEC, op. cit. note 57; F. Yang, D. Xin, and M.D. Levine, *The Role of Cogeneration in China's Energy System* (Berkeley, CA: Lawrence Berkeley Laboratory, 1997); K. Capoor, A. Deutz, and K. Ramakrishna, "Toward Practical Implementation of Article 4.1 of the Climate Treaty," Paper prepared for Climate Change Analysis Workshop, Springfield, VA, 6–7 June 1996; Walter V. Reid and Jose Goldemberg, *Are Developing Countries Already Doing as Much as Industrial Countries to Slow Climate Change?* (Washington, D.C.: WRI, July 1997).

90. Watson, et al., op. cit. note 32.

91. Ibid; Merlyn McKenzie-Hedger, "Agriculture and Forestry: Identification of options for net GHG reduction," Working Paper 7, Annex I Expert Group on the UN FCCC (Paris: OECD, July 1996); R. Neil Sampson, "Forests and global warming," *Journal of Commerce*, 2 July 1997.

92. Watson et al., op. cit. note 32; Anne Simon Moffat, "Resurgent Forests Can Be Greenhouse Gas Sponges," *Science*, 18 July 1997; McKenzie-Hedger, op. cit. note 91.

93. Watson et al., op. cit. note 32; McKenzie-Hedger, op. cit. note 91.

94. Ibid; Watson et al., op. cit. note 32; U.S. Department of State, op. cit. note 54; CNE and US CAN, op. cit. note 35; NAICC, op. cit. note 54; Environment Canada, *Canada's Second National Report on Climate Change: Actions to Meet Commitments Under the United Nations Framework Convention on Climate Change* (Ottawa, May 1997).

95. McKenzie-Hedger, op. cit. note 91; CNE and US CAN, op. cit. note 35.

96. Watson, op. cit. note 32; McKenzie-Hedger, op. cit. note 91; U.N. Framework Convention, "Report on...Germany," op. cit. note 53.

97. McKenzie-Hedger, op. cit. note 91; Ministry of Housing, op. cit. note 40; McKenzie-Hedger, op. cit. note 91; U.N. Framework Convention, "Report on...the Netherlands," op. cit. note 40.

98. Lovejoy, op. cit. note 45; Watson et al., op. cit. note 32; Worldwatch estimate based on Boden et al., op. cit. note 1, and BP, op. cit. note 1.

99. Watson et al., op. cit. note 32; "Planned and ongoing AIJ projects," *Joint Implementation Quarterly*, September 1997; "First Carbon Emissions Trading Transaction Occurs Between Costa Rica, U.S. Financial Firm," *International Environment Reporter*, 28 May 1997; "Norway First to Sign Declaration of Intent on World Bank Fund to Cut CO_2 Emissions," *International Environment Reporter*, 11 June 1997; IIEC, *Opportunity Knocks: The Export Market for the Energy Efficiency and Renewable Energy Industry* (Washington, D.C., 1996).

100. Caspar Henderson, "Right climate for change," *Financial Times*, 6 August 1997; IEA, op. cit. note 33; Lovejoy, op. cit. note 45.

101. Houghton et al., op. cit. note 1; Watson et al., op. cit. note 2.

102. Houghton et al., op. cit. note 1; Watson et al., op. cit. note 2; U.S. Department of State, op. cit. note 54.

103. Houghton et al., op. cit. note 1; Jerry Melillo, "Ecological and Climatic Consequences of Human-Induced Changes in the Global Nitrogen Balance," U.S. Global Change Research Program Seminar, Washington, D.C., 14 April 1997; Watson et al., op. cit. note 32; Watson et al., op. cit. note 2; U.K. Secretary of State for the Environment, op. cit. note 34.

104. Houghton et al., op. cit. note 1.

105. Elizabeth Cook, *Lifetime Commitments: Why Climate Policy-makers Can't Afford to Overlook Fully Fluorinated Compounds* (Washington, D.C.: WRI, February 1995); U.S. Department of State, op. cit. note 54; CNE and US CAN, op. cit. note 35.

106. Cook, op. cit. note 105; Ministry of Housing, op. cit. note 40; U.N. Framework Convention, "Report on...Denmark," op. cit. note 40; U.K. Secretary of State for the Environment, op. cit. note 34; "Fluorocarbon Producers Back Ban on HFCs in Drink Cans," *Europe Environment*, 8 July 1997.

107. U.N. Framework Convention, "Report on...Japan," op. cit. note 53; "MITI Plans to Regulate HFCs, Asks Partners to Do Same," *International Environment Reporter*, 9 July 1997; "MITI To Move Toward Development of 3rd Generation Freon," *Nihon Kezai Shimbun*, 11 August 1997; "Chipmakers to Jointly Tackle Substitute CFC Problems," *Asia Pulse*, 23 July 1997.

108. Hilary F. French, "Learning from the Ozone Experience," in Lester R. Brown et al., *State of the World 1997* (New York: W.W. Norton & Company, 1997).

109. John T. Houghton et al., *An Introduction to Simple Climate Models used in the IPCC Second Assessment Report*, IPCC Technical Paper 2 (Geneva, February 1997).

110. John T. Houghton et al., *Stabilization of Atmospheric Greenhouse Gases: Physical, Biological and Socio-economic Implications*, IPCC Technical Paper 3 (Geneva, February 1997).

111. Table 8 is Worldwatch ranking based on a scale of 1 to 4 in six policy areas: fossil fuel and road subsidy removal and carbon and energy taxes, energy efficiency standards, industry covenants, incentives for renewable energy and cogeneration, support for carbon sequestration, and measures to address non-CO_2 greenhouse gases; European Commission (EC), "Climate Change—The EU Approach for Kyoto," Communication From the Commission To The Council, the European Parliament, The Economic and Social Committee And The Committee Of The Regions (Brussels, 1 October 1997).

112. Anderson et al., op. cit. note 83.

113. Ibid.

114. French, op. cit. note 108.

115. EC, op. cit. note 111; Toru Kunimatsu, "Gov't sets 5% greenhouse gas reduction target," *Yomiuri Shimbun*, 2 October 1997; White House, "President Clinton Announces the United States Climate Change Policy," National Geographic Society, Washington, D.C., 22 October 1997; "Australian Diplomats Urged to Debunk US Climate Forecasts," *Sydney Morning Herald*, 28 August 1997.

116. U.N. Framework Convention, "Review of the Implementation of the Convention and of Decisions of the First Session of the Conference of the Parties," Second compilation and synthesis of the first national communication from Annex I Parties, Report by the Secretariat (Bonn, 27 June 1996).

117. French, op. cit. note 108.

118. Robert Repetto and Jonathan Lash, "Planetary Roulette: Gambling with the Climate," *Foreign Policy*, Fall 1997; Government of Japan, "GREEN Initiative: Global Remedy for the Environment and Energy Use," distributed at the Special Session of the U.N. General Assembly, New York, June 1997; Christopher Flavin, "Banking Against the Greenhouse," *World Watch*, November/December 1997; IIEC, "Financing Energy Efficiency in Countries with Economies in Transition," Working Paper 6, Annex I Expert Group on the UN FCCC (Paris: OECD, July 1996).

119. U.N. Conference on Trade and Development, *Controlling Carbon Dioxide Emissions: The Tradeable Permit System* (Geneva, 1995); Watson et al., op. cit. note 32; Fiona Mullins and Richard Baron, "International GHG Emission Trading," Working Paper 9, Annex I Expert Group on the UN FCCC (Paris: OECD, March 1997); Robert W. Hahn and Robert N. Stavins, "Trading in Greenhouse Permits: A Critical Examination of Design and

Implementation Issues," in Lee, op. cit. note 18; John J. Fialka, "Clear Skies Are Goal As Pollution Is Turned Into A Commodity," *Wall Street Journal*, 3 October 1997.

120. Edward A. Parson, *Joint Implementation and its Alternatives: Choosing Systems to Distribute Global Emissions Abatement and Finance*, Belfer Center for Science & International Affairs, Environment and Natural Resources Program, John F. Kennedy School of Government (Cambridge, MA: Harvard University, August 1997); Mullins and Baron, op. cit. note 119; Anderson et al., op. cit. note 83.

121. International Council for Local Environmental Initiatives, *Local Government Implementation of Climate Protection, interim report to the Conference of the Parties* (Toronto, July 1997).

Worldwatch Papers

No. of Copies

Worldwatch Papers by Christopher Flavin

_____138. **Rising Sun, Gathering Winds: Policies to Stabilize the Climate and Strengthen Economies** by Christopher Flavin and Seth Dunn

_____130. **Climate of Hope: New Strategies for Stabilizing the World's Atmosphere** by Christopher Flavin and Odil Tunali

_____119. **Powering the Future: Blueprint for a Sustainable Electricity Industry** by Christopher Flavin and Nicholas Lenssen

_____100. **Beyond the Petroleum Age: Designing a Solar Economy** by Christopher Flavin and Nicholas Lenssen

---

_____137. **Small Arms, Big Impact: The Next Challenge of Disarmament** by Michael Renner

_____136. **The Agricultural Link: How Environmental Deterioration Could Disrupt Economic Progress** by Lester R. Brown

_____135. **Recycling Organic Waste: From Urban Pollutant to Farm Resource** by Gary Gardner

_____134. **Getting the Signals Right: Tax Reform to Protect the Environment and the Economy** by David Malin Roodman

_____133. **Paying the Piper: Subsidies, Politics, and the Environment** by David Malin Roodman

_____132. **Dividing the Waters: Food Security, Ecosystem Health, and the New Politics of Scarcity** by Sandra Postel

_____131. **Shrinking Fields: Cropland Loss in a World of Eight Billion** by Gary Gardner

_____129. **Infecting Ourselves: How Environmental and Social Disruptions Trigger Disease** by Anne E. Platt

_____128. **Imperiled Waters, Impoverished Future: The Decline of Freshwater Ecosystems** by Janet N. Abramovitz

_____127. **Eco-Justice: Linking Human Rights and the Environment** by Aaron Sachs

_____126. **Partnership for the Planet: An Environmental Agenda for the United Nations** by Hilary F. French

_____125. **The Hour of Departure: Forces That Create Refugees and Migrants** by Hal Kane

_____124. **A Building Revolution: How Ecology and Health Concerns Are Transforming Construction** by David Malin Roodman and Nicholas Lenssen

_____123. **High Priorities: Conserving Mountain Ecosystems and Cultures** by Derek Denniston

_____122. **Budgeting for Disarmament: The Costs of War and Peace** by Michael Renner

_____121. **The Next Efficiency Revolution: Creating a Sustainable Materials Economy** by John E. Young and Aaron Sachs

_____120. **Net Loss: Fish, Jobs, and the Marine Environment** by Peter Weber

_____118. **Back on Track: The Global Rail Revival** by Marcia D. Lowe

_____117. **Saving the Forests: What Will It Take?** by Alan Thein Durning

_____116. **Abandoned Seas: Reversing the Decline of the Oceans** by Peter Weber

_____115. **Global Network: Computers in a Sustainable Society** by John E. Young

_____114. **Critical Juncture: The Future of Peacekeeping** by Michael Renner

_____113. **Costly Tradeoffs: Reconciling Trade and the Environment** by Hilary F. French

_____111. **Empowering Development: The New Energy Equation** by Nicholas Lenssen

_____110. **Gender Bias: Roadblock to Sustainable Development** by Jodi L. Jacobson

_____106. **Nuclear Waste: The Problem That Won't Go Away** by Nicholas Lenssen

_____105. **Shaping Cities: The Environmental and Human Dimensions** by Marcia D. Lowe

_____104. **Jobs in a Sustainable Economy** by Michael Renner

_____102. **Women's Reproductive Health: The Silent Emergency** by Jodi L. Jacobson

_____97. **The Global Politics of Abortion** by Jodi L. Jacobson

_____**Total copies (transfer number to order form on next page)**

W138

PUBLICATION ORDER FORM

_____ ***State of the World:* $13.95**
The annual book used by journalists, activists, scholars, and policymakers worldwide to get a clear picture of the environmental problems we face.

_____ **Worldwatch Library: $30.00 (international subscribers $45)**
Receive *State of the World* and all six Worldwatch Papers as they are released during the calendar year.

_____ ***Vital Signs:* $12.00**
The book of trends that are shaping our future in easy to read graph and table format, with a brief commentary on each trend.

_____ **WORLD WATCH magazine subscription: $20.00 (international airmail $35.00)**
Stay abreast of global environmental trends and issues with our award-winning, eminently readable bimonthly magazine.

_____ **Worldwatch Database Disk Subscription: $89.00**
Contains global agricultural, energy, economic, environmental, social, and military indicators from all current Worldwatch publications including this Paper. Includes a mid-year update, and *Vital Signs* and *State of the World* as they are published.
Can be used with Lotus 1-2-3, Quattro Pro, Excel, SuperCalc and many other spreadsheets. **Check one:** ____ **IBM-compatible or** ____ **Macintosh**

_____ **Worldwatch Papers—See complete list on previou[illegible]**
Single copy: $5.00 • 2–5: $4.00 ea. • 6–20: $3.00 ea. • 2[illegible]r more: $2.00 ea. (Call Director of Communication, at (202) 452-19[illegible], for discounts on larger orders.)

$4.00 Shipping and Handling *($8.00 outside North America)*

_____ **TOTAL**

Make check payable to Worldwatch Institute
1776 Massachusetts Ave., NW, Washington, DC 20036-1904 USA

Enclosed is my check or purchase order for U.S. $________

☐ AMEX ☐ VISA ☐ MasterCard ______________________________
Card Number Expiration Date

____
name **daytime phone #**

____
address

____
city **state** **zip/country**

phone: (202) 452-1999 fax: (202) 296-7365 e-mail: wwpub@worldwatch.org

website: www.worldwatch.org

HD9502.A2 F59 1997
128641
Flavin, Chris
Rising sun, gathering wind
3 0094 0004 9781 5